Praise for *How to Talk to Your Kids About Climate Change*

In her illuminating new book *How to Talk to Your Kids About Climate Change*, Harriet Shugarman provides a comprehensive description of the climate crisis, how to communicate it to the children in your life, and how you can work together to be part of its solutions. It's work like this book that makes me so incredibly inspired by, and proud of, our global network of Climate Reality Leaders.

—Al Gore

With all the love and wisdom of a true mother, Harriet takes you by the hand and fearlessly leads you to face the stark truth about climate change. Once there she doesn't let go, but rather uses that love to encourage us to step forward and become part of the solution. Harriet's vast experience, knowledge, and care for our planet and our humanity, vibrates through every page, giving us all the tools we need to arm ourselves and our children who are literally facing the fight for their future.

—Linus Roache, Golden Globe-nominated actor, and climate activist

This book is essential reading. As the global climate emergency imperils our present and future lives, climate education is vitally important. Read this insightful guide to why and how we can "talk climate" with kids!

—Raffi Cavoukian, C.M., O.B.C., singer, author,
Raffi Foundation for Child Honouring

Raising kids in a world made uncertain by climate change is a challenging task. Read Harriet Shugarman's book, *How to Talk to Your Kids about Climate Change*, to learn how we can speak to our children about climate change in a way that conveys urgency but maintains agency in the battle to preserve a hospitable future for us and them.

—Michael E. Mann, distinguished Professor,
Penn State University,
co-author, *The Madhouse Effect*

How to Talk to Your Kids About Climate Change is a precious and necessary book delivered at the perfect moment. Harriet Shugarman skillfully guides us through this most perilous time by helping us to look squarely at the climate crisis while simultaneously showing us how to navigate through the fire. Her words are all at once fierce and soothing, responding to many questions parents around the world seek to answer as their children look them in the eye.

—Osprey Orielle Lake, Executive Director,
Women's Earth and Climate Action
Network (WECAN) International

This book will sustain you in your commitment to protect, empower, and support your children to *act* at this defining moment of our planetary history. Indeed, taking action, together as a family, is the antidote to fear and demoralization.

—Lynne Cherry, founder, Young Voices for the Planet,
producer, Young Voices for the Planet films,
author and Illustrator, *The Great Kapok Tree, A River Ran Wild*,
co-author, *How We Know What We Know About Our Changing Climate*

As a mother, *How to Talk to Your Kids About Climate Change* lifts up my spirit, at the same time as it reminds me that our children are demanding the truth about our climate emergency, and that they are looking to us to help them understand the realities we are facing and the role they can play. Harriet Shugarman has written a must-read guide for parents at all stages of climate knowledge. I trust her to help me find new ways to talk to my children, to tell them the truth, and to empower them to move forward in positive ways. You should too! After reading this book, when your kids ask you about the climate crisis and what they can do about it, you will know what to say.

—Maya van Rossum, Leader, the Delaware Riverkeeper Network,
Founder, the national *Green Amendment For The Generations* movement,
author, *The Green Amendment*

Harriet is an iconic leader and founding member of the climate and parenting nexus. Her book expertly navigates the complexities of climate science, solutions, and the emotions that come with facing our global climate crisis. Parents and non-parents alike will greatly benefit from Harriet's personal journey exploring the hopes, fears, and realities of the state of our planet.

—Tara DePorte, founder, The Human Impacts Institute

Shugarman reminds us that we can't know how it will all turn out. Yet, we can find ways of not being alone and ways to navigate this climate emergency. There are steps parents can take to cope, and more importantly, cultivate perseverance in their children.

—Lindsay Coulter, David Suzuki's Queen of Green

From the brutal realities of the mounting threat of climate disruption comes this extraordinary work—brilliantly conceived and full of the inspiring messages we all need to hear in these fateful times. *How to Talk to your Kids about Climate Change* acknowledges our vulnerabilit—but expertly redirects us to the connections we must make with Mother Earth to restore her life-giving bounty. Sustaining the resolve to do this requires steely commitment to the job ahead—and to infusing the effort with purpose, love, and authenticity. Harriet has brought us a clear-eyed plan to navigate the journey. Keep this guide at your side—in her hands even the most treacherous shoals can become opportunities.

—Lise Van Susteren, M.D. Board Certified:
General and Forensic Psychiatry

With five grandsons, I appreciate Harriet Shugarman's honest and sensitive advice to parents and grandparents. Harriet has led the way outreaching to parents. We must unite to end carbon pollution while helping our children prepare for the changes already baked into the climate system. Harriet shows the way.

—Larry J. Schweiger, author, *Climate Crisis and Corrupt Politics* and *Last Chance*, former President National Wildlife Federation

As we live our climate emergency, the impacts of the crisis are not being felt equally. Being a parent is a delicate dance – preparing and at the same time protecting our children. I speak honestly with my son about climate change and show him through my actions how we all have a role to play. In *How to Talk to Your Kids about Climate Change*, Harriet shares messages of hope, the importance of telling the truth, of taking action, of climate justice, and of facing our fears head on. Read this book.

—Ambassador Thilmeeza Hussain,
Permanent Representative of the Maldives to the United Nations

How to Talk to Your Kids About Climate Change

TURNING ANGST INTO ACTION

Harriet Shugarman

Cover design by Diane McIntosh. Cover image: ©iStock

Printed in Canada. First printing May, 2020.

Inquiries regarding requests to reprint all or part of *How to Talk to Your Kids About Climate Change* should be addressed to New Society Publishers at the address below. To order directly from the publishers, please call toll-free (North America) 1-800-567-6772, or order online at www.newsociety.com

Any other inquiries can be directed by mail to

New Society Publishers
P.O. Box 189, Gabriola Island, BC V0R 1X0, Canada
(250) 247-9737

LIBRARY AND ARCHIVES CANADA CATALOGUING IN PUBLICATION

Title: How to talk to your kids about climate change : turning angst into action / Harriet Shugarman.

Names: Shugarman, Harriet, 1961- author.

Description: Includes bibliographical references and index.

Identifiers: Canadiana (print) 2020017536X | Canadiana (ebook) 20200175378 | ISBN 9780865719361 (softcover) | ISBN 9781550927290 (PDF) | ISBN 9781771423250 (EPUB)

Subjects: LCSH: Children and the environment. | LCSH: Global warming. | LCSH: Climatic changes. | LCSH: Child rearing.

Classification: LCC QC981.8.G56 S58 2020 | DDC 363.738/74—dc23

New Society Publishers' mission is to publish books that contribute in fundamental ways to building an ecologically sustainable and just society, and to do so with the least possible impact on the environment, in a manner that models this vision.

MIX
Paper from responsible sources
FSC
www.fsc.org
FSC® C016245

Contents

Acknowledgments

Special thanks to the following Climate Mamas and Papas for sharing their thoughts, their hopes, their concerns, and their advice with me. I am honored to have the opportunity to share their wisdom with you. Please note that ages of children, place of employment (where included), and state or city of residence, etc., are current only as of the writing of this book.

Olena Alec, Director of Engagement, The Climate Reality Project, Washington DC; child under age 1

Rachel Arkus, New York, NY; child age 3

Ana Baptista, Associate Director, Tishman Environment and Design Center (The New School), Newark, NJ; kids 7 and 11

Donna Bullock, Pennsylvania State Representative, Philadelphia, PA; kids 11 and 8

Lynne Cherry, Founder, Young Voices for the Planet, Maryland and Cape Cod

Helene and Raoul Costa de Beauregard, Founding members Parents For Future Seattle, Seattle, WA/France; kids 5 and 1

Anna Fahey, Director of Strategic Communication, Sightline Institute, Fidalgo Island, WA; kids 5 and 9

Ted Glick, progressive writer, activist, organizer, TedGlick.com, Bloomfield, NJ; adult child

Lorna Gold, academic, climate campaigner, and author, Ireland; kids 7 and 9

Derek Hoshiko, organizer, For the People, Clinton, WA; kids 2, 4, and 5, plus 8-year-old goddaughter

Larry Kraft, former executive director, Imatter Youth, Minneapolis, MN; kids 12 and 14

Jill Kubit, co-founder and director, DearTomorrow; co-founder, Our Kids' Climate, Brooklyn, NY; child age 6

Jill MacIntyre Witt, author, activist, Climate Reality Mentor and Leader, Bellingham, WA; two adult children

Allan Margolin, Creator of I ❤ Climate Scientists, Climate Nexus Hot News, ABC Political Producer, New York City, NY; kids 14, 12, and 12.

Diane McEachern, award-winning entrepreneur and best-selling author, Washington, DC; adult children

Kathy Mohr-Almeida, author, educator, journalist, and psychotherapist, Tucson, Arizona; child age 16

Eve Mosher, artist working on climate change, mother of Fridays for Future Youth, New York, NY; kids 9 and 7

Monica Mayotte, Boca Raton City Councilwoman, Boca Raton, FL; adult children

Jim Poyser, award winning journalist, fiction writer, essayist, Earth Charter Indiana, Indianapolis, IN; adult children

Perry Sheffield, pediatrician, researcher on environmental exposures at Mount Sinai, New York, NY; child age 5

Chrysula Winegar, nonprofit communications executive, Darien, CT; kids 17, 15, 13 and 11

To Osprey Orielle Lake—my mentor, teacher, and dear friend: thank you for always being there to listen, to share advice, and to hold my hand through hope and grief. To Angela Fox, thank you for bringing me fully into the anti-fracking community. To my Climate Reality family, too numerous to name names, as I will surely forget someone: you know who you are and how much you have supported and helped me along this lifelong journey we are traveling together. To Al Gore, thank you for having the vision, the knowledge, and the perseverance to create and to continue to grow our Climate Reality family, and for giving us all the tools we need to help build active hope.

Special thanks to the many people at New Society Publishers who helped make this book a better one. Thanks in particular to Rob West for your suggestions, ideas, and support and for your trust in me to find ways to make these happen. To Linda Glass, thank you for making sure that all the "i's" are dotted and the "t's" crossed, and for ensuring that the meaning behind each word is clear to all.

Last, but certainly not least and with so much love, thank you to my husband Andrew. I would not be a Climate Mama without you. Nor would this book ever have been written without your love and support. Thank you for being my confidant and best friend and for making everything a possibility. Thank you for always encouraging me to follow my passion, to pursue my life's work, and to discover new and varied ways to turn my *angst into action*.

Dedication

My Dearest Elliot and Alana,

For the first time in many years I feel that we are truly on the cusp of creating a livable future—one where you will not only survive but where you can thrive. A seemingly unstoppable wave of international and national climate action has begun; it is broad-based in scope—youth-led, empowering, and hopeful. Young people like you and even kids much younger are at the center, building this active hope, and parents stand behind, beside, in front—wherever needed. Your future is not yet written; we still have time. But I am fully conscious of the fact that we are running out of time. Mother Nature is demanding our attention, and she won't let us look away.

I love our family discussions about the internet of things, driverless cars, artificial intelligence, political movements, space travel, history, and the future. Many of our talks from years past about what the future will look like are already our reality. Yet, even as fast as life is moving, I feel that we are still in the driver's seat; we are needed to drive the car. But I also know our grip on the steering wheel is more and more tenuous. This moment of feeling in direct control will soon pass, both literally and figuratively. The steering wheel itself will soon disappear; the writing is on the wall. We must therefore drive fast and furious and with great intention while we still can.

I remain optimistic, resolved, and determined. I stare down my fears, and I will myself to be forever hopeful. When doubt begins to creep in, I banish it, and I think of you. I refuse to believe and will not accept that your future is predetermined.

I stand with you—wherever and whenever needed. I believe in your future and in each of you.

I recently reread a different letter[1] that I wrote to you when you were both in high school.

It reminded me that I must *keep* speaking out and yet at the same time be accepting of a wide range of approaches; there are many paths each one of us can take. Rereading that letter also reaffirmed that when we have the same end goal—a future where you will not only survive but thrive—we are infinitely stronger, even if we take different roads to get there. The voices, votes, hands, and levers of power that are needed to transform the ripples we are creating grow each day.

A tidal wave is coming.

Epigraph

Anything else you're interested in is not going to happen
if you can't breathe the air and drink the water.
Don't sit this one out. Do something.
You are by accident of fate alive at an absolutely
critical moment in the history of our planet.

—Carl Sagan

I have learned in my life that optimism is not the result
of changing the world but the route to doing so.
The coming years are a generational chance
to make a positive impact. We must see that chance
with a deep sense of the huge privilege that it is to be alive today
and a fierce, stubborn optimism.

—Christiana Figueres

Future generations may well have occasion to ask themselves,
"What were our parents thinking?
Why didn't they wake up when they had a chance?"
We have to hear that question from them, now.

—Al Gore

Preface: This Book is for You

This book is for parents of newborns, first graders, middle schoolers,high school seniors, college students, and everyone in between. It is also for parents of children who themselves are adults, perhaps even with children of their own. This book is for you as you come to grips with the dire realities we face. Raising a child takes a village. Even if you are not a parent, you are part of the village. Thank you for reading this book and for not looking away. Thank you for seeking out answers and for deciding to jump in with both feet—or at least one foot—while you think about the next step for foot number two.

My hope is that this book will inspire and empower you to look for more information, to learn as much as you can, and to become actively involved—through a myriad of ways that make sense to you. By exciting your passions, your hopes, and your desires for a future in which our children not only survive but will thrive, you will keep going, do more, stay committed, and be resolved to make change. We must do all we can to ensure a livable and hopeful future for our children and for us. Our actions start today.

What This Book Does

This book will provide you with the science, the reality, the words, the ideas, and the support you are looking for as you build paths to navigate and live our climate emergency. It will also help you come to terms with the grief that accepting our climate crisis brings, while at the same time, it builds active hope and a shared life preserver we can all hold onto—with room for whoever wants to grab on.

This book will help you talk to the children in your life. It offers a wide range of ways to show you care, that you are taking action on their behalf, and that you hear their concerns and are working on ways to put the brakes on—to slow down and help change the collision course we are currently on. This book is written from my perspective; it includes my biases and is based on my background and experiences. It isn't the definitive answer; nor is it the only way forward. It is, however, my best and most honest advice, gained from years of experience and from what I have seen, heard, and witnessed.

This book is not a list of or a resource of the *top ten things you can do to save the planet*. Those books abound. Find one, read it, and take action. None of the ideas you encounter here will be *rocket science*; nor are they untried, new, or revolutionary ideas—you shouldn't expect them to be. Many of the ideas relate to viable and constructive solutions that you likely are already employing or have thought about instituting.

Nor is this book a statement or a position on the politics or integrity of the United States under the current administration, under administrations before this one, or under administrations to come. We must remember that the lack of cohesive and urgent action on the climate crisis didn't begin with the Trump administration. As of the writing of this book, the climate crisis has *never* been acted on with the overriding urgency that our government or other national governments around the world have the power to command. There remains plenty of blame to go around.[1]

What is becoming more evident as technology advances, as climate modeling becomes more precise, and as our planet reels from her inabilities to absorb and counteract the impacts of our assaults on her, is the reality of our direct role in causing our climate to change. Each year, as we learn we have met or exceeded another temperature, drought, rain, snow, or "natural" disaster record, humankind's direct role in changing Earth's climate and the ensuing climate chaos that is unfolding is becoming more evident and more apparent.

This book is my way of bearing witness to what is happening around us. Watching, sharing and witnessing our evolving climate crisis—full time—is my job. Whether the boulder moves up the hill, or sometimes slowly rolls back down, we are living climate change; this is our new normal.

Timing is Everything

The 2019, ten-year anniversary of ClimateMama, an organization I founded in 2009 to help parents learn more about our climate crisis, provided an opportune moment for me to stop and take stock, to look at where we have been, where we are headed, and where I think we should already be. The stories shared with you in the following pages, unless otherwise stated, are ones seen through my eyes. They are my experiences, told in my voice. They are my truth and my reality, unvarnished, as I have seen and experienced them.

This book also includes the voices, thoughts, ideas and advice of Climate Mamas and Papas from across the country and around the world; these are people I look up to, admire and trust. They all work in one way or another on finding solutions to our climate emergency—in support of and for their children and yours. I am honored and thrilled that they have shared their experiences and thoughts with me, so that I can share them with you.

On the occasion of my wedding, a friend who had traveled from afar to be with us told me she came because she wanted to be my witness to this milestone in my life. This book is my opportunity to bear witness in a public way, to share what I see, what I have seen, and what I have learned. It is also just a snapshot, a moment in time. As I have been watching, I have witnessed the clear acceleration of our climate crisis. When you read this, the facts themselves will be that much more obvious—as evidence of our climate crisis and our hand in causing it mounts and accumulates daily. As of 2018, with scientific reports giving a 2030 closing window for the ability to slow down our climate crisis, and with Greta Thunberg's youth climate strikes giving rise to an international youth uprising, public outcry for immediate action is accelerating. Oddly, I think that advice about how we speak with and listen to our children's hopes and concerns about our climate crisis will remain stable even as the seriousness of the crisis comes more sharply into focus.

I would suggest to you, as I often do to myself, to stop—even just for a few minutes—to meditate, to think deeply and to take stock of what is happening around us. Sit still and silent, let yourself feel the depth and weight of the crisis we have created, even as you recognize and accept how painful and difficult this is. Taking off your rose-colored glasses and truly seeing and understanding, even if only for a few moments, the enormity of the changes we, as a species, have imposed on our natural world is both terribly sad and awe inspiring; at the same time, it is incredibly humbling. We need to recognize and acknowledge what scientists have been telling us for decades with increasing authority. The destructive planetary forces that we—as a singular species—have been able to unleash in an incredibly short time span, are unique in the history of our planet.

There are deniers out there who continue to say we are too small and insignificant to make or create a planetary impact. Lately, there are also doomers on the other extreme; those who tell us it's too late, that we are out of time, and that the damage we have wrought will destroy us all. The two groups are on opposite sides of the spectrum, but they are both trying to paralyze us. However, the vast majority of climate scientists—the experts who study what is happening and who are showing us we, indeed, are the ones changing our planet and our climate—are still encouraging us to act. They are reminding us we still have time: the door is still open. But at the same time, these climate scientists—as well as Mother Nature and our children—are working overtime to warn us that the door is spring loaded and closing rapidly. As protectors of our children, we are called to act now. We are the ones who are truly woke, and once awake, it's impossible to ever truly sleep soundly again.

Coming to terms with our climate reality requires that we constantly move forward and act with the urgency our crisis demands. This doesn't, however, mean neglecting to live life or failing to stop and smell the roses. After all, these roses have flourished because of our family experiences and the memories we are making with and for our children. For many of us, our children are why we are determined to go further and farther than we think we can. They are why we refuse to look away—why we are working so hard to find ways to accept and understand what is happening now, even as we force ourselves to acknowledge what will come if we do not act.

Our future and the future for our children can therefore seem very bleak. There is no denying this. But, it can also be filled with endless opportunities and new ways to become engaged, to take action and to get involved. We can choose to trust in our actions and those of millions of others who are working on and toward climate solutions, even as we recognize we can't know for sure how things will turn out.

We must imagine the future we want, envisioning our hopes and our dreams. We must envision a future where climate change is a chronic condition, one we have figured out how to live with, one that is manageable and allows our children to thrive, not just survive. Our children are watching us; their present and future is truly in our hands. Let's talk honestly with them about this—about the endless opportunities, about the active hope.

Let's begin by telling our children the truth.

1

Introduction

Awakening to Our Climate Emergency

We will be coming to terms with the urgent realities of our climate crisis for the rest of our lives. We all need to get comfortable with this fact—in a very uncomfortable way. (Easier said than done…)

Each morning for a moment, as I gaze intently at my sleeping children resting in blissful peace, I am refilled with resolve and hope. I remind myself that it is my job to secure a safe and livable future for them and to ensure that they have the opportunity to grow into adults—to fight for their future as I now fight for my own and for theirs. Yet a game of chance is underway, with my children's future the ultimate prize. Our house is literally on fire, from the Amazon to the Arctic Circle. We can smell the smoke and we can see the flames. There is still time to put the fire out, to save at least some of our belongings and maybe even the house itself, but we must act now. We are out of time.

Why won't everyone wake up? The fire is burning hotter and hotter.[1]

Dropping the Climate Change Bomb

I promise not to sugarcoat the truth. I will write as honestly and directly as I can. But, as David Wallace-Wells begins his book, *The Uninhabitable Earth*, "It is worse, much worse than you think."[2] Wallace-Wells gives us the worst-case scenario: what will happen if we do nothing to curb our greenhouse gas emissions. Many feel Wallace-Wells's assessment is fair; as of 2018, carbon emissions are climbing worldwide. It is an honest and clear assumption that we are headed in the wrong direction.[3] Yet in most media markets around the world, the climate crisis gets little to no sustained attention, and the emergency we face is regularly downplayed. From my vantage point, I can see clearly that we are not yet on the path that leads to a livable future for our children. However, I do feel strongly that the path forward is visible and becoming clearer; it will be hard work, but it is within our reach.

I ask myself constantly: will our current and future actions be enough to spur our transformation, staving off the chaos that now seems close to inevitable? I honestly don't know. No one can say for certain. There are so many forces working against us. At the top of the list is our own inability to fully comprehend the dire emergency we are actually living. Considering the money and power working overtime to keep us quiet and paralyzed, it is no wonder we remain mostly numb to the news that *our sky is falling*. Also, the reality is that we are often simply too overwhelmed by the demands of our everyday lives to dig in and work on solutions. Too many of us feel there isn't anything that we, as individuals, can actually do that will effectively hold up our piece of the sky. As a result, we often tend to second guess what nature and science are telling us.

Doubt creeps in as we ask ourselves

- Are things really so bad?
- Climate chaos will impact my kids—really?
- Aren't these problems still many generations away?

If *we* feel this way—unsure and doubtful—how will we be able to guide our children to find ways they can feel positive about their own future? How do we tell them the truth about our climate crisis and help move them to action when we aren't sure how to rally ourselves? Our climate crisis—a crisis of our own making—is relentlessly enveloping our lives, our homes, our children's health, our food systems, and our clean water resources. This is happening *right now*.

The two profound questions I believe we must honestly ask one another, and in all likelihood have already asked ourselves, are these:

1. Will we be able to cope, to adapt, to survive in the midst of this chaotic climate onslaught?
2. How will we maintain the strength and the determination to move forward and live our lives fully and with purpose as we live climate change and face the realities of each passing day?

The answers to these two questions are not clear; nor are they straightforward. Yet we need to come to terms with the answers if we are to be effective in talking to our children about climate change. A strong beginning is to acknowledge that living climate chaos—being alive in the Anthropocene[4]—is our reality. In certain places at different times each year, as extreme weather events wreak havoc, we know things have changed and we are living a new normal. Yet, at other times, as weather seems *normal* and reminds us of days from our youth, this new normal seems less certain. But we mustn't fool ourselves or pretend we aren't currently living a climate emergency. We *are*.

Why We Should Be Concerned

Since the year 2000, scientists around the world have begun using the term *Anthropocene* to describe our current geological time period: *Anthropo*, for man, and *cene*, for new. Not only did we enter a new millennium, we are in a new epoch. While there is still debate in the scientific world about the exact start date, more and more scientists feel strongly that we have left the Holocene, the epoch that our planet occupied since the last ice age, and we now have entered—whether we are ready for it or not—the age of man.

We humans are now more powerful than hurricanes, volcanoes, and tornadoes. We are able to change and alter the Earth's ecosystems. We have forced our planet out of balance, and we are causing her to spin out of control. A single living organism now threatens the survival of almost all species on Earth—and that organism is us. This is unprecedented in the history of our planet.

I was born in the 1960s. The level of carbon dioxide in the atmosphere was already being tracked. It was evident to scientists that these levels were beginning to climb. At approximately 320 parts per million (ppm), the level of CO_2 when I was born, we were still well within a safe range, and not too far from where the level of CO_2 had been for over 800,000 years.

When I was young, we had terrible wildfires, downpours that flooded towns and cities, nor'easters, and hurricanes. New heat and cold records were occasionally set. But, our *new normal* of extreme weather events is of an entirely new dimension and scale. Almost monthly, heat records are not only being broken, they are being shattered. Nighttime low temperatures are no longer as low as they have been historically, and nighttime high temperatures are setting new highs. New high temperature records are being established around the world with regularity; these are then broken and reset again.

Climate scientists tell us that the safe limit for carbon dioxide in the atmosphere for our planet (and for us) is around 350 parts per million (ppm). Over the past 50 years, we have contributed to an extraordinary rise in the level of carbon dioxide in the atmosphere, surpassing 415 ppm in 2019[5]—a level not seen on our planet in millions of years. For most of us, 350 parts out of a million seems much too small to alter our entire planet's atmosphere. Parts per million? We passed 400 ppm? What's the big deal? Yet, scientists tell us that any more than 350 ppm becomes the magic elixir—or, more appropriately, the poison apple—that puts our planet's atmosphere out of balance and can lead to conditions in which most species alive today, including our own, may no longer be able to exist.

I often see people's eyes gloss over or wander around the room when I share these startling scientific findings. But for so many reasons, and in particular because carbon dioxide levels track historically with temperature, we need to pay attention, to care, and to understand the importance of this data, which is irrefutable. "*Using sedimentary records and plant fossils, researchers have found that temperatures near the South Pole were about 20°C higher than now in the Pliocene epoch, from 5.3m to 2.6m years ago*...this was the last time that scientists believe that CO_2 levels in our atmosphere were as high as they are now."[6] At that time, our planet was ice-free and sea levels were hundreds of feet higher than they are today. Our planet is working overtime to maintain some form of control as we create changes that took thousands of millennia to occur in the past. But scientists tell us that temperatures could, in fact, rise quickly and suddenly too. They also regularly remind us that our oceans, which are also working overtime to absorb much of this excess heat and CO_2, are reaching their saturation point. We need to be concerned. We need to be outraged. We need to act now.

Also alarming and disturbing is that CO_2 isn't the only human-made greenhouse gas that is trapping heat in our atmosphere. Methane, which is another name for *natural gas* is also on the rise; so, too, are nitrous oxide, HFCs (hydrofluorocarbons), and other human-made short-term climate pollutants like black carbon and water vapor. As these gases collect in our atmosphere (some for very short periods of time), they just might be enough to push us, and our planet, over the proverbial climate cliff. Scientists tell us we will know we are there when we reach the point where we see increasingly and consistently wild weather events, dramatic sea-level rise that will destroy coastal cities and be accompanied by food and water shortages, the proliferation of vector-borne diseases taking flight across the globe, the death of coral reefs and old growth forests, and extreme air pollution blanketing major cities. And these are just a few of the miseries we might be facing. Yes, it does sound like a horror movie or the Book of Revelations, but we are on track for this to be our reality; it has already begun.

I wish we could go back in time. I think 30 years or so would be enough. I would travel back to the United Nations Earth Summit that was held in Rio de Janeiro, Brazil, in 1992. At this gathering of world leaders, a global spotlight was shone directly on the climate crisis. This was unprecedented; it was the first time that world leaders jointly and publicly recognized the realities and urgency of our climate crisis. World leaders pledged to work together as an international community to fix the problems we faced. They even outlined a step-by-step plan of action. As nation-states, we were already working together on solving an international environmental crisis. The hole in the ozone layer was identified in the late 1980s, and by the early 1990s—through international cooperation, joint planning and aggressive global action—the hole was already beginning to close. The crisis of climate change, while much more complex, was still something we believed we could and would be able to tackle. Programs and policies that would slow down the expanding crisis in real and concrete ways were possible and were being discussed. Coming together to devise and implement these plans seemed eminently doable.

If only.

If only the world community had jumped in with the urgency that was demanded at that time, we would be in a much more hopeful and very different place than we are today.

If only we hadn't been so naive about our direct and pervasive role in the unfolding crisis. We talked a good talk in 1992, but we certainly didn't walk the walk that the evidence—even back then—demanded of us. There is no do-over on this one. What we must do now is move forward—this time with our eyes open wider.

In the fall of 2018, climate scientists, through the United Nations Intergovernmental Panel on Climate Change (IPCC), told world leaders in their *Special Report: Global Warming of 1.5°C—Summary for Policymakers*[7] that we effectively need a World War II-style mobilization to drastically reduce our climate pollution, or we will face runaway climate chaos as soon as 2030. At the time this IPCC report was released, the date of *no return* was only 12 years away.

This is our reality.

Clearly, this is no longer the stuff of science fiction, Hollywood movies, or something that's happening way *over there*. Our daily climate news is as relentless as it is devastating. Yet for a moment in time in late 2018, the directness of this United Nations report seemed shocking enough to wake up a huge number of people around the world, at the same time. The IPCC report appeared on the front pages of the *New York Times*, *The Washington Post*, *The Guardian*, *USA Today*, *The Globe and Mail*, and many other prestigious papers across the globe. Even people in the climate and environmental community who had been warning about this emergency for years were stunned to see climate chaos spelled out so clearly. The hope now was that, with so many others finally awake, collective climate action commensurate with the scale of the alarm being raised would soon follow.

If only.

Unfortunately but not surprisingly, the realities of this report, as well as previous and subsequent ones, seem to be regularly wiped off the front pages almost as quickly as they appear—buried and out of sight within days of their release. Sadly, today's headlines are eerily similar to headlines that have come and gone each year since the Earth Summit of 1992. What does change though, is the growing *urgency* of our climate crisis, backed up with more and more peer-reviewed scientific data. Mother Nature also regularly chimes in with her own increasingly harsh warnings.

As I write this book in 2019, I think how quickly the years fly by. Twelve years ago, my daughter was 8 and my son was 9. Just a few years before that, we had moved from New York City to New Jersey, and ClimateMama was not yet even a dream. Although I feel like I have hardly aged since that moment, I look at my grown children and the wrinkles on my face that mark the passage of time. Twelve years ago, my father was alive and the cancer that killed him was invisible, perhaps lurking or maybe not even there. I recently renewed my driver's license; it now expires in 10 years, yet my photo on the license will remain the same. Twelve years seems like a lifetime ago, and yet it seems like it was only yesterday. Time marches on, and it is not on our side.

I hope that 12 years from now, my children will have careers they love and significant others in their lives who love them as much as I do, and that we all have our health and happiness. I imagine them living long and full lives in which they grow old gracefully, fulfilled in their careers and happy in their life choices. I close my eyes and see this future with every fiber in my body. In this future, our world community has found effective ways to come together; we have transitioned off fossil fuels; we have created a circular economy and jobs for all; and, collectively, we have brought this great and just transformation to fruition. I imagine the future I want, and this pushes my fears down and out of sight.

Regrettably, the facts facing us from our growing climate crisis keep mounting, and our window for slowing things down—for giving our planet time to heal and for creating hope for a livable future for our species—seems to be closing. With each passing year that we don't take clear and definitive action, this window closes more quickly.

The Solutions Puzzle

We are all part of the solutions puzzle—a puzzle with an infinite number of pieces—a puzzle that keeps us intrigued, excited, and motivated to keep adding to it, as we work diligently to complete it. Each piece has a defined place. Whether it's one piece at a time or a series of pieces put in place together, each piece moves us closer to a more complete picture and our primary goal: *a livable present and a livable future.*

Yet, most of us realize (whether consciously or subconsciously) that this puzzle will never truly be completed—the final pieces are unknown, and some are already out of our grasp. We all must find ways to work together on this puzzle. Those of us already understanding the depth of our crisis must make working on the puzzle our constant passion, our challenge, our opportunity, and our enduring hope. The rest of us can help by simply identifying and collecting like pieces, and occasionally finding ways to fit pieces together, so that we help lessen the load of those working full time on the puzzle.

Even now, as newly minted grownups (Alana is 20 and Elliot 21), my children love working on puzzles. This has been a family tradition since they were preschoolers, and it remains an ongoing side activity over family holidays and vacations. Whether the puzzles are easy or hard, my children will often work together for hours and sometimes even over days and weeks. But always, they do so knowing there is an end in sight; the box tells them how many pieces are included, and it shows them the completed goal. Would they even begin the puzzle if they knew all the pieces weren't there? Doubtful.

Climate reality is not a labyrinth. Climate scientists tell us there is still a path through the climate chaos that lies before us, around us, and ahead of us. It may not be clear or obvious, and at times it will (and does) feel impossible, but the way forward does exist. We must keep piecing the puzzle. Our children are counting on us.

In order to keep going, we must be unafraid to look truth in the eye. There is no one magic seed that will grow the beanstalk to help us reach the castle high in the clouds. Instead, there are billions of seeds, overflowing handfuls for every person on our planet. Each one of us is part of the larger solution, a piece of the puzzle of climate hope. Some of the pieces fit easily together; others are much more challenging to find. Even once we find them, we may still have difficulty seeing how they fit together.

Similarly, we all have different abilities, degrees of opportunity, and differentiated powers with which to make change. We all can and must do what we can. This is the magic, the trick, the opportunity. Those that can go big, must. As Canadian artist and activist Franke James has said, "We must do the hardest thing first."[8] Often, this is the most difficult step to come to terms with. If you are reading this, then clearly you are willing and interested in learning how.

Remind your children of these four facts:

- What is happening is NOT normal.
- The crisis unfolding is still within our control.
- We can slow down the changes that are taking place.
- The crisis doesn't rest on their shoulders alone.

Putting our heads in the sand, hoping it will all go away or that someone else will fix it, won't work. We can look to the left, to the right, behind us, and in front of us, but what we need to do is look directly in the mirror: we all are part of the solution. We need to begin in earnest to adapt, to build up our resiliency, and to create new ways and new opportunities that will give our children a chance at a livable world—one that won't be managed by and through chaos. We must continue moving forward by leaps and bounds but also one step at a time.

Let the ClimateMama motto help guide you:

> **Tell the truth.**
> **Actions speak louder than words.**
> **Don't be afraid.**

Internalize these intentions as well as the power of *many ripples*. I have done so, and my climate hope remains active and is constantly renewed. Know that there is no one answer or one path forward; there are billions—at least as many paths as there are people on our planet. Letting our passions guide us is the way forward. These are the lessons we can and must teach our children.

I alone cannot change the world,
but I can cast a stone across the waters
to create many ripples.

—Mother Teresa

Climate Awakening: Climate Reality

Clearly, I am not the first parent to wake up and feel the uncomfortable and heavy weight that our growing climate crisis puts on our children's present and future. You, too, may also feel these same weighty feelings, or you may be just waking up to them. Others will continue to join us. However, what is becoming more evident with each passing year is that those of us alive today may be the last generation to have the ability to slow down the runaway train that is our climate crisis. We need to slow down the train to give our children and others the chance to jump on and join us, because only together can we *act on climate*; and this we must do, for the rest of our lives.

My climate awakening came in the summer of 2006; I remember it well. I can place the time and the date. My children were in grade school; I had recently moved out of New York City; and my full-time and then my part-time day jobs working at the United Nations had finally come to an end.

On the surface, things looked fine, but inside I felt confused, unsure and untethered. Like others who'd made the move from the big city to the 'burbs, I was finding it difficult to comfortably settle into my new suburban life. What would be my next steps, both personally and career-wise? I worried whether I would ever figure this out, and if I would ever be completely comfortable in my new setting. I felt like I was *disappearing*, and I wasn't sure what would make me reappear.

I had left a really big stage in New York City and was now bringing up my own children—in a suspiciously similar setting to the one I had been raised in. Similar to my own upbringing, my family's house in the suburbs had a front and a backyard and a public school and community playground just down the street. There was a big city a short distance away, yet the city seemed a world apart from the everyday hum and routine of home. My young family, similar to my own growing up, was fortunate and advantaged on so many levels. I didn't fully grasp this, though, at that point in my life.

While my husband and I had grown up in different countries (he in New Jersey and I in Alberta, Canada), in many ways our neighborhoods and our upbringings were eerily similar. Growing up as we did in the '60s and '70s, the world seemed to hold many possibilities. In actual fact, we were both born less than 20 years after the end of the Second World War. Yet, the war and all its grave and earth-shattering damages were somehow positioned for both of us as if they were ancient history. And certainly, if the Second World War was ancient history, slavery and colonization, with all their brutal components and legacies, seemed like relics of an ancient and long bygone past. Both in suburban Canada and suburban New Jersey, our lives and our home lives were stable, with seemingly endless opportunities.

Clearly, we both were extremely privileged to have grown up where and how we did—a fact neither of us fully appreciated. I now understand that being born in certain zip codes (US) or postal codes (Canada) is a clear marker of the level and quality of many things: education, health, food security, and ultimately even the job opportunities that will unfold over a person's lifetime. This remains true today, and in many ways it is even more pronounced than it was when I was growing up. Today, we continue to hide in plain sight dangerous infrastructure we don't want to see, such as power plants, incinerators, garbage dumps, and chemical plants. We hide this infrastructure in marginalized neighborhoods, often those belonging to immigrants and poor people of color. Our climate crisis is already impacting these neighborhoods first and worst, but it will ultimately upend even the "tidy" more economically advantaged neighborhoods, too. No one can or will stay safe, hidden, or untouched by our unfolding climate crisis.

In 2006, the movie *An Inconvenient Truth* opened in theaters across the US, and the public's consciousness of the implications of global warming was ignited. I knew then that my life would be forever changed. After seeing the movie, my eyes were opened wide, and I wouldn't let them shut. I felt that I had a role and a responsibility to help others open their own eyes—to make them understand the extent of the climate crisis we had set in motion. I began to look for ways to put my passion into action. Like many others, I wanted to be involved in something bigger than myself, and here it was, right in front of me.

Eight months after I saw the movie *An Inconvenient Truth*, I was in Nashville, Tennessee, at one of the first group trainings of the newly formed Climate Project, now known as The Climate Realty Project.[9] Around a table in a Nashville hotel ballroom, I found myself chatting and sharing my concerns and hopes with people who—on the surface—seemed very different from me: an ordained minister, an elected municipal official from the South, an NFL football player, and a climate scientist. Yet I quickly came to realize that these people were actually very similar to me; they understood intimately what I was thinking and feeling without my having to explain it.

The 150 or so amazing people filling that Nashville hotel ballroom came from every corner of the country and every walk of life. They all came to Nashville ready, willing, and able to take a stand and make their voices heard. We were all in agreement: it was up to us to raise the alarm. Our future demanded it. Strength in numbers all of a sudden had a direct and powerful meaning for me. I felt surer of my ability to actually create change. I had worked on climate issues on the international stage for many years, yet, somehow, looking directly at the climate crisis at that moment—and through my parent lens—activated me in ways I hadn't been awake to before. Knowing that each person I was with would return home and be embedded in their local community, working on climate action, made me feel more positive and seemed more direct to me than the work I had been doing at the United Nations.

My marching orders on leaving Nashville were crystal clear: *educate, empower, and activate my community on the climate crisis*.

A day or two after I returned, I was helping out at a special event at my children's school. I saw hundreds of single-use plastic water bottles being used and then thrown directly into the regular garbage—there was no formal recycling program at my children's school. I hadn't paid attention to this before. I also noticed parents idling in their cars as they waited to pick up their children from school. These two observations presented a way in. These were direct opportunities for change that I could help create.

I recognized that these would be small changes, but if I helped put in place a recycling program and a no-idling policy at my children's school, I would be reaching hundreds of families, a way of being more involved locally. These small steps following my Climate Reality training have now grown into a full-time career working on the climate crisis and a life-long commitment to building and creating climate action.

In the ensuing years, I have traveled far. I have become a nationally recognized spokesperson on our climate crisis. I was arrested in Washington, DC, standing strong for my beliefs. I have spoken at national business meetings and conferences, on state house steps, at rallies, in schools and church basements, in packed auditoriums across the country, in intimate gatherings in kitchens, and in the living rooms of friends. I have been featured on national television, in YouTube videos, at an off-Broadway theater production, on radio and on podcasts. I write for national and regional publications. I am a college professor, a community leader, an organizer and an activist. I have helped shape government energy policy and organized events to bring the reality and truth about the climate crisis to people around the country.

I have to admit, though, that at times I have felt like an imposter. Are people listening to me? Am I really making a difference? I was told there is a name for this. It is known as imposter syndrome. Have you ever had this feeling? I imagine we all feel this way at times, questioning ourselves, our abilities, and our actions. But I have come to this conclusion: Why *not* me? I advise you to do the same. We must each do what we can and then some. I am a proud and at times fierce Climate Mama, with two children who deserve a chance for a happy and livable future. I have the ability, the passion, and the opportunity to work toward this future for them—nothing can stop me.

I have stayed actively involved with my Climate Reality family and with the Climate Reality organization that helped launch the trajectory of my climate-focused career. I have come to recognize that my work on the climate crisis is the great work of my life. I have been a mentor at Climate Reality trainings and have served as a district manager across three states. I established the first local chapter for Climate Reality, the Climate Reality Project New York City Metro Chapter, and I am the Chapter's first chairperson. In March 2017, in Denver, Colorado, I welcomed more than 1,200 new leaders to our family from the stage, following Al Gore and former Colorado governor, Bill Ritter. Twice, I have been a guest on the multimedia, international broadcast of *24 Hours of Reality*; and over the years I have represented Climate Reality at events around the country.

In October 2017, when I was a mentor at the Pittsburgh, PA, Climate Reality training, 1,400 new climate leaders joined our family in three days; this was one of our largest trainings to that point. More people were trained in those three days than had been trained in the six events we held in the first year of trainings in 2007! At our Pittsburgh training, I was surprised, humbled and honored when I was awarded the prestigious Green Ring, given *to an outstanding Climate Reality Leader who has demonstrated an exceptional commitment to their role as a climate communicator and activist*. This followed a profile earlier that year of me in Al Gore's 2017 book *An Inconvenient Sequel: Truth to Power*. In August 2019, I joined Mr. Gore on stage at our Minneapolis training to share my story of becoming and being a Climate Reality leader. While I work with many other climate-oriented organizations that are doing wonderful work, my Climate Reality family is where I feel most at home. *An Inconvenient Truth* launched my travels from Mama to ClimateMama, inspiring me and forever lighting a burning passion and powerful spark in my lifelong journey of education, exploration and boundless climate hope.

ClimateMama Beginnings

ClimateMama was started with little fanfare in March 2009. I registered the website name, found a web developer to work with, and began writing some initial blog posts and the first pages of the website. We had a second soft launch in October 2009.

The website introduction I wrote in 2009 remains relevant today in almost all respects. Minor updates have included changing the phrase *climate change* to *climate crisis* and lessening the onus on individual actions by placing more emphasis on collective action, policy action, and direct action. Everything old is new again. Yet, as climate scientists and Mother Nature continue to issue dire warnings about the vulnerability of our species, those who could put in motion giant leaps remain slow to act. We must shine a light on these organizations and individuals, demanding their full attention and that they take immediate action now.

ClimateMama Website Introduction: Welcome to ClimateMama

You are a mother, a father, a grandparent, an uncle, an aunt, a teacher, or a child at heart. When you hear the Native American saying, "We don't inherit the Earth from our ancestors, we borrow it from our children," it makes you stop for a moment and think. You love nature, travel, adventure, and believing in a world that is special and unique. *Climate change* and *global warming* are words that alarm you, that often seem too big to get your arms around. You care about what's happening to the world and notice small changes in your own life that seem to point in the direction of a threatened environment. But you wonder if these changes are real, and if they are, you can't imagine what you can do to help change what is happening.

ClimateMama will give you the facts on Climate Change. We will give it to you straight, in a clear and easy format to understand, from trusted and reliable sources that have been vetted and triple checked. We want you to understand clearly what climate change and global warming mean, and what they mean to you, to your day-to-day life, and to the lives of the ones you love. We want to keep you updated on the latest developments around our planet, on what is happening now, and on what is important for you to know. We want you to meet ordinary people who are doing extraordinary things to fight this incredible challenge we face. We want to help you choose products and services you can feel good about, that will help us all reduce our carbon footprint and help the Earth heal itself. We want to help you understand what your carbon footprint is and give you advice on fitting into a smaller size. We want you to be able to look your son, your daughter, your grandchild, nephew, niece, student, or yourself in the eye and say with certainty that you are doing what you can to take care of the Earth. We want you to feel empowered, that you are doing what you can to ensure that the world our children inherit is a world that is as precious, as vibrant, as healthy, and as alive as the one you grew up in.

Let's all make changes in our daily lives that affect climate change so we can each play our part in helping to solve this difficult puzzle before us.

Throughout the first year of our ClimateMama launch, I would check our analytics daily, sometimes multiple times a day—getting excited when two or three people who weren't from New York, New Jersey, or Edmonton, Canada (my original hometown), perused our website. I would often try to figure out who the person was who opened our website, and I think I was regularly successful in doing so! Our readers now number in the tens of thousands, and I no longer wonder. I am overjoyed to see that so many people around the world find our articles, ideas, and thoughts useful and thought provoking.

Memorable Blog Posts

Over the years, I have written hundreds of blog posts for ClimateMama and many other articles and posts for a wide range of local, regional, national, and international publications. For me, writing these posts has awakened mixed feelings of joy, hope, and sorrow, and I am told regularly by our Climate Mamas and Papas, that it has done the same for them too. To give you a flavor of these posts, I share two with you now.

"Climate Change & Cancer: The Long Goodbye"[10] is one of my most reflective posts. I began writing it in 2015, the year of my father's death, and I shared it for the first time on the Elephant Journal website in 2016 on my first Father's Day without my father. It comes from the deepest of places; it is a story for and about my father Maurice, and it is directly from my heart. Below is a reflection on that post that I wrote on my *second* Father's Day without my dad.

ClimateMama blog, June 16, 2017

Father's Day Without a Father

Father's Day 2017 is my second Father's Day without my father. Life happens, we all lose loved ones—this is a fact and the circle of life. While these heartbreaking losses stay with us forever—the crushing, immediate, and paralyzing hurt does slowly fade.

In 2016 I wrote a post that was published in Elephant Journal, called "Climate Change & Cancer: The Long Goodbye." The post drew parallels—using my personal experiences and thoughts—between the campaign to keep cigarettes in people's hands and today's climate denial. By downplaying the clear links that 100s of peer-reviewed scientific studies showed between cigarette smoking and cancer, many people became addicted to cigarettes and experienced negative health impacts—including death. Many articles and studies have shown how the same tactics used to downplay linkages between cigarette smoking and cancer are now playing out to cast doubt on our role in accelerating climate change. A small group of well-funded climate deniers has been raising doubt about the linkages between our use of fossil fuels and our heavy footprints on our planet and the exponential acceleration of climate change that threatens our species and all other species as well. These climate denier campaigns have been brewing for years, with success.

"I am still raging inside, shaking with anger. Raging against the disease that killed my father and raging against the political machinery that lets too many of our political candidates continue to publicly, vocally, and with authority deny the realities of climate change, raising doubt about our role in what is happening and thereby slowing down solutions and actions.... Our planet is showing us in a myriad of ways that it is critically sick and suffering. And every day, new scientific studies connect the dots between our way of life, our energy choices, our actions, and our changing climate—yet these studies are not translated into the necessary tough policies we need set us on a safe path." My feelings last year, when things were still so raw for me on so many levels.

This year, my rage has subsided to a slow simmer, and my thoughts on this Father's Day have moved to reflection and memories—of times well spent with my father and on building and growing my resolve. I feel strangely hopeful that we have turned a corner; our cards and those of our current administration are on the table. In many circles we are moving to stronger and lasting solutions on climate change, in others, still farther away.

No, I haven't drunk any strange Kool-Aid, and yes, I do understand that we are at the edge of a cliff, but I feel our momentum has shifted to resolute and sustained action both in the United States and around the world. Here in the United States, it seems clear to me that we are publicly and visibly being asked to take sides, to look our children in their eyes and tell them where we stand. If this isn't obvious to all, then we must make it so. All our elected officials, and each of us, must be clear on where we stand. Will we join with countries, cities, states, businesses, colleges, and organizations around the world, who are loudly declaring they are "still in" moving forward with firm, immediate, and sustained climate action—or not. From where I sit, the "moving forward side" is picking up speed and seems unstoppable.

Donald Trump and many in his Cabinet and inner circle have clearly and vocally staked their allegiance to coal, fossil fuels, big agriculture, and limited environmental regulations, coupled with a clear and loud amping up of the climate denial machine. But I remain hopeful because each day I see pushback to this message and how this approach is being marginalized and side stepped. Clearly our democracy, and the place of the federal government in it, is being seriously challenged, but it has been for some time. What is more obvious to me is that climate change policy has more overtly become a partisan issue and divide. This is NOT okay, and more and more people are pushing back. Now more than ever we must stand together with friends and family—whether, green, purple, blue, red, Republican, Democrat, or independent—and declare loudly that climate change is real, here, now, and caused by us, and yes, that we CAN do something about it.

My father's death from cancer was directly linked to his years as a smoker—to be sure, exacerbated by other conditions. Understanding the connections between cigarettes and cancer didn't make it any less painful for our family or for my father—who showed us all how to live with dignity, honor, and hope, even with the knowledge that death was knocking loudly at the door. As my father and our family grappled with the disease that killed him, we had our eyes wide open to the cause, but we did pin our hopes on prayer and on treatments not quite there. Prayer helped some of us feel stronger, and now, almost two years after my father's death, the immunotherapy drugs that might have extended his life are available and being widely used. Things can and do change quickly.

Climate change in its current manifestation is a problem of our making; one that has unfolded over years and will take years and lifetimes to resolve. But as we debate reality, and as climate denial is allowed to ferment, we are wasting time we can't get back. Clearly, in this world of "Alternative Facts" post-election, telling the truth has been put into question and the disease we have inflicted on our planet has a death grip on our human species. We know that this hasn't happened overnight. And the easy thing would be to place blame solely on the current administration. But this isn't an easy problem where we can point a finger to identify the cause; nor is there one or even a handful of solutions. There ARE a myriad of solutions, with room for us all to be involved.

So, my rambling message to all our Climate Mamas and Papas this Father's Day, as we come together to celebrate our fathers, our brothers, our grandfathers—living and dead—is that we must stay hopeful, be reflective, and be intentional. My father showed through his actions how to live life this way. I strive each day to honor his memory and to follow his example.

Happy Father's Day, Dad... I love you, I miss you and I think about you every day.

The second blog post I want to highlight was published the day before the 2016 presidential election and less than a month shy of the one-year anniversary of the historic Paris Climate Agreement. World leaders were preparing to gather in Marrakech, Morocco, for the annual United Nations Conference of the Parties,[11] on the dawn of the 2016 US presidential election. The intent of the United Nations Morocco meeting was to show the world how the international community would *walk the walk* on climate change policy and actions. The election of President Trump effectively put a pin in that international balloon of climate hope. The wisdom shared in this ClimateMama post remains valid today. I share it now because the sentiments shared in it are important to revisit. We must constantly remind ourselves that fighting for climate justice and a just transition to a renewable future will take time but that this future is being built carefully, on a solid foundation.

Interesting aside: while researching the original 2011 blog post (this 2016 post updates a 2011 post) and googling "famous climate quotes," the vast majority of quotes found on climate change, the environment, Earth Day, and global warming were from men. This fact remained predominately true in 2016; but it is changing rapidly today. Many of the loudest voices we are hearing demanding climate action are the voices of young women and girls. We need diverse voices speaking out on our climate crisis. Our kids, regardless of gender orientation, ethnic background, or place of birth should be able to see themselves as climate leaders; and it seems they are.

ClimateMama blog, November 7, 2016
Eight Climate Quotes from Inspiring Women

In 2011, ClimateMama published one of our most widely read posts: "Top 10 Climate Change Quotes from Inspiring Women." I thought it would be illuminating to reach out to some of these same inspiring women and see if and how their thoughts on our climate crisis had changed and evolved. We reached out to them for comments in October of 2016, one month prior to the 2016 US presidential election. By all accounts, we were on the cusp of electing a woman to the highest office in the land—this, following eight years of a black president filling that office. Anything seemed possible.

I was excited to share the optimism, hope, and urgent call to action of these eight incredible women—ClimateMama superstars—who were and still are taking action and moving forward on climate solutions—always leading, never following. Their quotes are remarkably thoughtful and thought provoking even to this day.

Grumbling about injustices, wrongs and frustrations doesn't make anything just or better—however, sharing matters! Paying mindful attention and converting outrage into meaningful action in order to bring about positive change really works. Do something each day to make things better for yourself and for others, because we are in fact, consciously or not, making a difference with every action, and every choice we make—one way or another. I live and work by this African Proverb—*If you think you're too small to make a difference, try sleeping in a room with a mosquito.* It is incumbent upon each of us—as individuals—to share more, to involve more people, and to leave a healthier world for our children.

—Lisa Borden, founder Borden Communications, strategist and catalyst, enthusiastic philanthropist, inspiration agent, and wannabe organic farmer

When I imagine my future grandkids asking me what I did to fight climate change, I will be able to say I did 'something'.... I will tell them we did the hardest thing first by selling our only car, an SUV. I will tell them that we laughed when people called us "granola-crunching tree-hugging whack jobs." I will tell them that when I wrote to the Prime Minister asking him to make polluters pay, I never thought he would take notice... But then I will tell them that the government does read their mail. Behind-the-scenes government officials warned people not to show my art because, "The artist's work dealt mostly with climate change and was advocating a message that was contrary to the government's policies." It was scary to be told in effect, "Do Not Talk about Climate Change. It is against government policy."... Then I will tell my future grandkids that when people told me not to talk about climate change, I talked about it more!

—Franke James, artist and author focusing on free expression, human rights, disability rights, and our environment; Awarded the PEN Canada/Ken Filkow Prize in 2015 and the BCCLA Liberty Arts Award in 2014

If the next four years are spent rolling back whatever progress has been made on emissions, then almost certainly the temperature targets that world leaders set last year in Paris will be breached. In fact, even if the next four years are spent making more progress, it's likely that the targets will be breached. In the case of climate change, to borrow from Dr. King once again: "tomorrow really is today."

—Elizabeth Kolbert, staff writer for *The New Yorker*; Pulitzer Prize winner for "The Sixth Extinction: An Unnatural History," *The New Yorker*, September 2016

I am privileged to have the opportunity to work with youth and teachers, witnessing regularly the innovative thinking and collective will needed to build more sustainable and just communities. We need full engagement, commitment and follow through on climate action on regional, national and international levels. Creating space for youth to authentically engage in this process will serve us all well."

—DR. LISA (DIZ) GLITHERO, professor and educational researcher, University of Ottawa; Canada C3 Education Lead

Humanity is currently enacting a narrative that nature is ours to abuse and exploit and pollute as we see fit, forgetting that we are a part of it. We are part of the web of life, and when we harm one part of that web, we harm ourselves. We urgently need a new narrative, where instead of hubris we have humility. Instead of rapacious destruction we have respect and stewardship. Instead of disconnection, we have deep connection—to nature, to each other, to ourselves, and to our future.

—ROZ SAVAGE, ocean rower, environmental activist and author; speaker at "Our Ocean Conference," Washington, DC, September 2016

There is no more powerful voice than that of a mother, father, caretaker, nurturer. We, as stewards of this planet, need to engage even more deeply to confront climate change and environmental injustice. We are the adults, the leaders—and must show our children that we are fighting in earnest for their future, and for the collective future of humankind."

—LAUREN SULLIVAN, director and co-founder of Reverb

Since I realized that global warming posed a threat to my children's—all children's—health and well-being, one of my most rewarding discoveries has been that our voices and votes really matter to our elected officials. I'd thought that we just needed to make lifestyle changes, like recycling. However, I learned that the scale and urgency of the climate crisis is so great that the only way to effectively tackle the problem is to change our polices. Our elected officials make those policies. Talking to our elected officials is an act of parental responsibility. Voting is an act of love."

—Dr. Sarah Warren, speaker, coach, and consultant, founder of Our Spheres of Influence

Throughout this book, I will continue to emphasize my firm belief that we must not let partisan politics get in the way of working together. We do not have the luxury of time on our side; we must all jump in with both feet. We must be crystal clear that whomever we entrust with our vote—at the local, regional, or national level—must be ready to leap forward on climate solutions.

I have had the privilege of meeting some of the most eminent climate scientists of our time—Kevin Trenberth, Katharine Hayhoe, Michael Mann, Jennifer Francis, Gavin Schmidt, and James Hansen among them. From my observations, they are all good and caring individuals—eminent scientists giving their all to help us understand this existential threat to our species. They and other climate scientists have committed themselves and their life's work to providing the hard data, the facts, the reality, so that there is no ambiguity, no *two sides* to a manufactured debate on what is real and what is just slowing down progress on climate action. Many climate scientists are also parents and grandparents, and, like you, they are concerned and worried about the world their children are inheriting.

As you learn more about the science of climate change in the next chapter and look closely at the issues of politics, fear, and justice—it's okay to feel mad, to be angry, and to let sadness creep in. I hope that as the facts become clearer and more evident to you, and your understanding grows, your resolve and your commitment to action will build as well. This should help you develop and hone your own ideas for talking to your children about climate change.

Our climate emergency is and will always be an emotional rollercoaster. As the emotions along the ride wash over you, let them;

it's important to feel and to acknowledge them. As with any roller-coaster ride, there is also exhilaration and excitement. As you take climate action on, your ride through life and the emotions that come when you know you are reaching people will be exhilarating. Let these emotions rise to the surface. Keep them there. And, as with any rollercoaster ride, there is also fear and sometimes despair and doubt about what is around the next corner; keep your eyes open as you round that next bend. Once you have reached the next straight-away, the path will once again become clearer, and you can let go of those feelings of concern. Allow resolve to set in as you steel yourself to round the next bend, drop, or steep rise. There is no getting off. You are on for the duration.

2

Thanks For the Ride, Dinosaurs

Science, Politics, Growth, and Justice

We have had an amazing ride on the backs of dinosaurs for about 200 years now, but it's clearly time to say thanks and get off. Our addiction to fossil fuels is literally driving us down the road to species extinction—our own included. We need to put the brakes on *now*.

We can point to the development and use of fossil fuels—coal, oil, and gas—as the spark that launched the Industrial Revolution, modern growth and development. Fossil fuels helped our species advance and progress at a pace unseen previously in human history. They propelled the United States, Canada, and most European countries to positions of power and modernization that benefited their citizenry and helped drive advances in science, technology, and medicine, helping people the world over.

However, science is now helping us understand and see the invisible atmospheric and down-to-earth damages and destruction that our reliance on fossil fuels has caused. Continuing the use of these fuels, unchecked, will drive climate chaos forward with a speed and pace also unseen previously in human history, or in the history of our planet, period. Digging deep into our planet's surface, dredging and extracting what time and Mother Earth have successfully buried for eons, has proven to be a deadly problem—certainly not the long-term solution first imagined.

And yet, it seems we can't let go.

Many governments, banks, pension funds, and stock markets around the world are continuing to find ways to subsidize and fund new and innovative methods of extending the lifeline of fossil fuels, ensuring their continued dominance and further expansion. When conventional drilling seemed to have reached its peak around the early 2000s, new forms of fossil fuel extraction were developed, including horizontal fracturing, or fracking. Fracking is a method of extracting oil or gas that involves extensive drilling—sometimes vertical miles underground and then miles horizontally. Chemicals and fluids are then forced down these drill holes, under pressure, fracturing the shale deposits that hold small deposits of oil and gas. These dregs—buried deep underground—are pulled out so that they themselves, and the other *travelers*[1] that come to the surface with them, can poison our water, our air, and us. In 2005, the US government created exemptions under the Clean Air Act and the Clean Water Act for oil and gas companies so that the fracking boom could expand in earnest. If these companies actually had to *clean up their messes*—something we teach our children from a young age—they would no longer be in business.

Another interesting way to account for our current mess is to consider attribution: who is actually creating the mess in the first place? According to the Climate Accountability Institute (CAI), *nearly two-thirds of carbon dioxide emitted since the 1750s can be traced to just 90 companies* that hail primarily from the fossil fuel and cement industries. The CAI research aggregates historical emissions to "carbon-producing entities." Rather than attributing emissions to nations—or individuals like you or me—their work identifies the direct source.[2]

So, while we agonize over our use of a plastic straw, reducing our meat consumption, or taking one less trip to see our children living on the opposite coast—90 companies producing, using and extracting fossil fuels are primarily responsible for changing our atmosphere and perpetuating our climate emergency. Clearly, this attribution method is only one way of partitioning blame. We are all addicted to fossil fuels; getting rid of their use isn't going to be simple or easy. Fossil fuels have permeated almost everything we do, and they are the building blocks of many of the items we use daily. However, if we can get even half of these 90 companies on board to work on climate solutions, cleaning up our mess seems that much more doable.

A 2019 report by the International Monetary Fund showed that worldwide subsidies for fossil fuels were in excess of $5.2 trillion—or over 6% of worldwide GDP.[3] The report states, "Efficient fossil fuel pricing in 2015 would have lowered global carbon emissions by 28 percent and fossil fuel air pollution deaths by 46 percent, and increased government revenue by 3.8 percent of GDP." This, as my kids used to say, "seems like a no brainer." If we just let the market dictate the price, the dominance of fossil fuels would begin a rapid decent. Yet fossil fuel subsidies actually *increased* from 2015 to 2017. Clearly, these large corporations don't have incentives to slow down or change what they are doing. Rather, *business as usual* is supported and encouraged.

The good news is that there is a direct role each of us can play in speeding up our transition to renewable energy and away from fossil fuels. For example, instead of voting for representatives who create incentives or subsidies that perpetuate our use of fossil fuels, or instead of putting your money in banks that fund or invest in fossil fuel development and its infrastructure, choose to support and elect representatives who are working on solutions to the climate crisis, and put your money into banks that are doing the same. Do your homework. Read your bank's annual report. Check your representatives' voting record, and ask them directly what their position is on fossil fuel use versus renewable energy. They need to have more than an opinion; they need to have a plan and a timetable for implementing it.

David Brower, founder of Friends of the Earth and co-founder of the League of Conservation Voters, is often called the father of the modern environmental movement. A New Year's Eve story he is credited with telling puts the history of our planet into just one short year. Reminding ourselves how fleeting our time on our planet has actually been, and the terrifyingly destructive reality of our accomplishments—enacted in the blink of an eye—puts in perspective how truly precarious this tightrope we are walking really is:

> January 1st marks the origin of Earth. By the end of February, the first simple cells appear. All the way through the spring and early summer, simple plants enrich the atmosphere with oxygen.
>
> Around mid-August, complex cells emerge, and coral appears in the ocean. Beginning in mid-November, the oceans fill with multicellular life-forms. In the last few days of November, freshwater fish appear, and the first vascular plants begin to grow on land.
>
> About December 1st, amphibians venture onto dry land. The great swamps that formed today's rich coal beds existed between December 5th and 7th. On December 12th the largest of the Earth's mass extinctions wipes out 95% of all species.
>
> Life bounces back, and dinosaurs evolve on December 13th. Flowering plants come on the scene on December 20th. In another great extinction, the dinosaurs disappear shortly before midnight on December 26th, opening a space for modern mammals to emerge on the 27th.
>
> On the evening of December 31—about when you might gather with friends for the evening's celebration—the first hominids evolve in East Africa.
>
> At 10 minutes to midnight on December 31st—about when all the New Year's party-goers are really starting to watch the clock—Neanderthals spread throughout Europe.
>
> At one minute to midnight, agriculture is invented. The Roman Empire fills 5 seconds, and collapses at 11:59:50—the moment when the New Year's ball starts to slide down the pole at Times Square, and the great 10-second countdown begins.
>
> In the last 2 seconds before midnight, we enter the modern industrial era. In those last two seconds we find the explosive growth of the human population, the rise of complex

technologies, and what we might call a globalized human culture."[4]

The entire history of the United States fits into the last second of this narrative. The "petroleum era" of cheap and plentiful energy is crammed into the last half of a second, as we're holding a deep breath, ready to shout our start-of-a-new-year greetings. The fireworks start as our dash through Earth's history brings us to the current moment.

Time flies.

Science

Climate crisis headlines, circa early 2000s:

- Giant African Baobab Trees Die Suddenly After Thousands of Years, Scientists Link to Climate Change
- The Great Barrier Reef Is Being Battered by Climate Change, More Than Half Has Bleached and Died
- Deaths in Puerto Rico from Hurricane Maria in 2017 Equal or Exceed Those That Died on 9/11
- Earth's 6th Mass Extinction of Species Has Begun, Also Threatening Global Food Supply
- Zika Spreads Silently Up and Down the Coasts of the United States
- The Growing Area for Coffee Beans Is Shrinking and the Coffee Plant's Existence Is Threatened
- Worldwide, Climate Scientists Agree That If Greenhouse Gas Emissions Are Not Dramatically Decreased by 2030, Irreversible and Catastrophic Changes to the Planet Will Be Locked into Place
- In 2019 US Midwest Farmers Lose Most of Their Crops to Unrelenting Rains; Food Security in Many Parts of the US and in Export Markets Around the World Faces Serious Challenges
- Beginning in 2019, Power Utilities in California Regularly Put in Place Rolling Brown Outs As Dry Conditions and Possible Electric Sparks Spread Fear of Devastating Wildfires

- More Than 600 Million Indians Are under Conditions of Extreme-Water Stress and 21 Indian Cities Are Predicted to Run out of Water by 2020

And these are just a few of the headlines I could list.

Scientists have known for more than 100 years, and fossil fuel companies have been aware for decades, about the linkages between the burning of fossil fuels and the creation of a chaotic climate system. When I inform students in my Climate Policy and World Sustainability classes that scientists began explaining the greenhouse effect almost 200 years ago, and that scientific analysis and early understanding of the impacts from human-created greenhouse gases goes back nearly as long, they are both surprised and shocked. They ask me if these facts are purposefully hidden. They question why this information is downplayed and kept out of the public consciousness. And they demand to know why, if we know what we know, we aren't taking drastic measures.

In 1965, Frank Ikard, then-president of the American Petroleum Institute, stated the following:

> One of the most important predictions of the report[5] is that carbon dioxide is being added to the Earth's atmosphere by the burning of coal, oil, and natural gas at such a rate that by the year 2000, the heat balance will be so modified as possibly to cause marked changes in climate beyond local or even national efforts.

Earth is sometimes called the *goldilocks* planet, or the *just right* planet. Our complex home is regulated and controlled by a climate system developed over billions of years that has fostered the growth and development of millions of amazing species, including our own. As of late, particularly over the past 200 years as we have begun amplifying nature's impacts by creating our own greenhouse gases, we have tipped this delicate balance, as more and more solar radiation is trapped within our atmosphere.

Notable scientific advancements leading to and explaining the linkages between climate change and human activities:

- **1824:** Joseph Fourier (France), mathematician, works on heat transfer; credited with earliest research and discovery of greenhouse effect.
- **1856:** Eunice Foote (US)[6] studies long-term changes in atmospheric carbon dioxide levels and their effect on temperature of the Earth. Her paper was presented at the 8th Annual Meeting of the American Association for the Advancement of Science.
- **1859:** John Tyndall (Ireland) conducts detailed experiments on radiative forces of certain gases. Shares proof with the world that greenhouse gases can impact the Earth's temperature. Tyndall is widely credited as the first to prove this theory.
- **1896:** Svante Arrhenius (Sweden) connects the dots between increases in carbon dioxide from human emissions in our atmosphere and climate change.
- **1938:** G.S. Callendar (England), engineer and inventor, links rising carbon dioxide to temperature rise and greenhouse effect, renewing scientific interest in connections to human impacts.
- **1957:** Roger Revelle (US) publishes paper that links increased carbon dioxide in the air to fossil fuel use. Dominant figure for many years on connecting these dots and recognizing their ramifications.
- **1958:** Charles David Keeling (US) developed and established recording instruments on the Hawaii Mauna Loa volcano, measuring CO_2 in the atmosphere. This remains the longest consecutive measurement of atmospheric CO_2, providing definitive proof of rising CO_2 levels.
- **1959:** Edward Teller (Hungary/US), physicist. In a speech at the 100th anniversary party of the American Oil industry, he stated,

> At present the carbon dioxide in the atmosphere has risen by 2 per cent over normal. By 1970 it will be perhaps 4 per cent, by 1980, 8 per cent, by 1990, 16 per cent [about 360 parts per million], if we keep on with our exponential rise in the use of purely conventional fuels. By that time, there will be a serious additional impediment for the radiation leaving the earth. Our planet will get a little warmer. It is

hard to say whether it will be 2 degrees Fahrenheit or only one or 5. But when the temperature does rise by a few degrees over the whole globe, there is a possibility that the icecaps will start melting and the level of the oceans will begin to rise.[7]

- **1965:** Edward Lorenz (US), early computer model developer working at the intersection of math and meteorology. He developed early models that looked at the chaotic nature of the climate system and the possibility of sudden shifts.
- **1969:** Mikhail Budyko (Russia) and William Sellers (US) in separate studies presented models of catastrophic ice-albedo feedbacks. Sellers wrote, "The major conclusions of the analysis are that man's increasing industrial activities may eventually lead to the elimination of the ice caps and a global climate much warmer than today."[8]
- **1972:** Club of Rome report, *Limits to Growth* identified and measured global challenges facing humanity, including limits to natural resources and the environment.
- **1975:** Independent studies find that CFCs and also methane and ozone can make a serious contribution to the greenhouse effect.
- **1977:** Exxon's scientists report to top executive management committee that emerging science shows carbon dioxide levels were rising—likely driven by fossil fuel use—and that such increases would increase global temperatures, leading to widespread damage.
- **1979:** US National Academy of Sciences report finds it highly credible that doubling CO_2 will bring about 1.5–4.5°C of global warming.
- **1985:** Expert consensus is reached during a conference in Villach, Austria, on the expanding depletion of the ozone layer and the unprecedented increases in greenhouse gases. Villach is seen as the first cohesive scientific call for governments to consider international agreements to restrict emissions.
- **1987:** Montreal Protocol of the Vienna Convention imposes international restrictions on emission of ozone-destroying gases.

- **1988:** James Hansen (US), NASA scientist, testifies before US Congress that, with 99% certainty, warming of Earth's atmosphere and buildup of greenhouse effect is caused by the buildup of carbon dioxide and other human-created greenhouse gases in the Earth's atmosphere.
- **1990:** The first report of the United Nations International Panel on Climate Change (IPCC) states that the planet has been warming, and future warming seems likely. Industry lobbyists and some scientists dispute the tentative conclusions.
- **1992:** The United Nations Environment and Development Conference, or Earth Summit, was the largest world gathering to that date of heads of state on the environment. It established international mechanisms and bodies to monitor and discuss the ongoing climate crisis.

Fossil fuel companies had actual knowledge of the risks of their products and had taken proactive steps to conceal their knowledge and discredit climate science while at the same time taking steps to protect their own assets from the impacts of climate change... beginning as far back as 1959.

—Rhode Island Senator SHELDON WHITEHOUSE, March 5, 2019

The science has been clear for decades. With each passing month and year, there are more and more peer-reviewed scientific studies that clearly put the blame for our climate crisis squarely on the shoulders of humankind. In the early decades of the 21st century, the definitive scientific proof seems to slowly have entered the wider public consciousness. At the same time, however, the crisis is accelerating with alarming speed.

Echoing my students, the question remains: "If we have had this knowledge for so long, why haven't we done something?" While this information can be found with ease now, it still remains largely ignored or simply not taught or reported. I remind my students that the internet and the ease of sharing data and information certainly didn't exist 200 years ago—or even 30 years ago. Web searches also come with the ability to choose the news and the reality we want. We are seeing science and facts questioned on a regular basis. Scientific research is being threatened and tied to a false narrative that our climate crisis isn't as urgent or as present as it really is.

Time flies.

In late 2018, about a month after the release of the IPCC report *Global Warming of 1.5°C*, the "Fourth National Climate Assessment" (NCA) was released. This report was prepared by 13 federal US agencies that make up the US Global Climate Change Research Program. At the request of Congress, these agencies report to Congress and the president every four years on the climate crisis. What both the IPCC report and the NCA tell us, in no uncertain terms, is this: *We are living climate change. It's real, happening and caused by us. It is really bad, and it will get worse. However, if we take serious action, starting now, we can slow things down.*

In large part, because of the burning of fossil fuels, we have *loaded the dice*, setting the stage for storms and extreme weather events to be much worse than they would have been in the absence of human-caused greenhouse gases. We have already witnessed many wildfires, floods, droughts, and other extreme weather events that are much more catastrophic than they have been in the past.

Gaia is a tough bitch—a system that has worked
for over three billion years without people.
This planet's surface and its atmosphere and environment
will continue to evolve long after people
and prejudice are gone.

—Lynn Margulis, biologist and proponent of the Gaia Theory, which says that Earth is a living organism

---

Now, when it rains, it pours; when fires burn, they burn hotter and with greater intensity; and when we have storms and hurricanes, they are stronger than storms were in the past.

Politics

Having worked at the United Nations for more than 13 years, I know the alphabet soup of acronyms for United Nations committees, conferences, departments and meetings. It's something you generally have to be inside the organization to fully grasp. Below, as reference and with brief explanations, is a summary of the main United Nations bodies established to oversee international work on climate policy.

The international process of managing and working toward a stable global climate system began in earnest at the United Nations Conference on Environment and Development (UNCED), sometimes also called the Rio Earth Summit, which was held in Brazil in 1992. The conference launched Agenda 21, a global action plan on sustainable development, and the Rio Declaration, which was created to clarify the rights and responsibilities of states on environment and development. In addition, the Statement of Forest Principles was adopted, creating a set of principles to guide the sustainable management of forests. This conference also formally established the United Nations Framework Convention on Climate Change (UNFCCC).[9] The primary goal of the UNFCCC is to prevent dangerous human interference with the climate system. Two other UN conventions were introduced at the 1992 Earth Summit, the Convention on Biological Diversity (CBD), and the Convention to Combat Desertification (CCD). All three of these bodies now have close to 200 national governments as members. Just as the Security Council was created as a forum where nations could come together to guard world peace and prevent war, these forums were created so that governments could come together to discuss critical issues impacting our planet.

The Conference of the Parties (COP) is the decision-making body of the UNFCCC; it meets annually. COP21, the 21st annual Conference of the Parties, was held in Paris, France, in late 2015.

The agreements reached at this COP are colloquially known as the Paris Agreement. This Agreement is widely quoted and touted as a major international achievement on climate. The Paris Agreement was the first formal international agreement since 1992 with near-unanimous recognition that climate change is real, happening, and serious; and with new concrete plans to combat the climate crisis, and halt and reverse greenhouse emissions.

However, it is important to note that the Paris Agreement was (and remains) non-binding; moral suasion and public opinion are the only enforcement mechanisms. Each country came to Paris with its own plan on how it would reduce its national greenhouse gas emissions. Some countries also shared climate adaptation goals as well. Each country established its own plan, with no international input. This country plan, formally called an Intended National Contribution (INDC), was and is meant to be tightened every five years, by each individual country. There is no formal enforcement mechanism in place to make this happen. Currently, the United States, which did sign the Paris Agreement, is the only country to indicate it plans to pull out of the Paris Agreement.

The broad goal of the Paris Agreement is to keep world temperatures below a 2°C increase above pre-industrial levels, with an aspirational goal of 1.5°C degrees. It remains well recognized that even 1.5°C will render many small island states uninhabitable because of sea level rise and will wreak havoc on the shores of many countries, devastating many of the world's largest cities.

The first formal stocktaking of progress on country goals, or INDCs, will take place in 2023 and every five years thereafter. Each signatory country to the Paris Agreement has committed to tighten their goals regularly. The Paris commitments, established in 2015, are not stringent enough to keep temperatures below a 2°C increase; much greater and faster reductions are required if the 2°C goal is to be met. Only time will tell if countries are willing and able to make these commitments.

The Paris Agreement was the first collective attempt by each country in the world—developed and developing—to publicly commit to significant greenhouse gas reductions. Like most countries', the United State's commitments under the Paris Agreement were modest, at best. President Obama knew he could not get a more significant agreement passed by Congress. President Trump has promised to remove the US from the Paris Agreement; immediately upon taking office, he began to roll back many of the Obama administration commitments under the Agreement. Formally, the US cannot leave the Paris Agreement until the day after the 2020 US elections. Interestingly enough, the US is on track to meet and even exceed its modest Paris commitments by the first stock taking in 2023. This is because many US states, cities, and corporations have stepped up to dramatically reduce their own emissions, keeping the US on the Paris commitment track.

Having spent many years at the United Nations, and now watching the US move to a more and more isolationist position, I clearly see the diminishing role and relevance of the United States on the world stage. Since the Second World War, the weight and power of the US as the world's first and most powerful superpower—both in military might and moral suasion—has been undeniable. But as the US, under President Trump, pulls our country into isolationist positions across many fronts, our control on the rudder and steering wheel of international climate policy is being overtaken by our own subnational governments and corporations, and by countries and regions beyond our borders. In the absence of strong US national leadership, this is a good thing. The world needs a strong international coalition to lead and prepare for what is ahead, a coalition led by national governments aggressively moving forward on climate goals and actions, where all parties are all in, with both feet.

From my vantage point as a new American (I became a US citizen in 2006), the hyper-partisan politics of the United States is more than a little overwhelming. Yet, when we look back into US history, hyper-partisanship isn't something new, nor is it something we should be surprised or shocked by. Democracies are fragile the world over, including in the United States. They need to be tended and cared for in strangely similar ways to how we would tend and care for a garden. Currently, in the United States we are not tending our garden, and it is becoming overgrown with weeds.

I didn't really understand how fragile the democracy in the US was or could be when I voted in my first presidential election in

2008. As I watched the first bi-racial president in US history come to power, I thought democracy ran deep and strong within the borders and boundaries of the United States of America. I listened intently to President Obama's inspiring and hopeful speeches, and anything and everything seemed possible, including finally addressing the climate crisis with the urgency that we knew was required.

Unless we free ourselves
from a dependence on these fossil fuels...
we are condemning future generations
to global catastrophe.

—BARACK OBAMA, 44th president of the United States

President Obama used variations of this quote in numerous speeches in the lead-up to his election in 2008. He spoke regularly about our climate crisis and the need to solve it. President Obama definitely talked the talk. His 2008 campaign slogan was "Hope and Change." *If only* the slogan had been "Hope and *Climate* Change!" Had President Obama tackled our climate crisis directly—and during his first term—we might be in a different situation than the one we find ourselves facing today. Instead, he took on our climate crisis in earnest only in the last two years of his second term. It seemed to me he finally began to walk the walk only at that point—even as a divisive Congress held one hand behind his back.

If only.

We can't go backward; we can only go forward. We can't undo the mistakes we made, but we can commit to not making the same mistakes in the future. And we need to stop looking for one solution and one answer. There isn't just one; nor is there a linear way to move forward. The scope of our climate crisis is too deep and the impacts too powerful. Hope and change can and must rest on a million ripples.

A primary target to be forcefully reckoned with continues to be the powerful fossil fuel industry and lobby. Curtailing this industry and then helping it, if possible, to see the light, so it can begin in earnest to transition to and help build up the renewable energy industry, also requires stopping the spread and growth of new fossil fuel infrastructure. This infrastructure fight came fully into national focus in the United States in 2011 with the No Keystone XL protests. The Keystone XL pipeline project is a proposal to build an oil pipeline from northern Alberta to the Gulf of Mexico. This pipeline would transport Canadian bitumen[10] from its source—the Alberta oil sands—to refineries in the southern US, where it would be processed and then sold on international markets, to be distributed around the world. This pipeline is still not completed.

The 2011 Keystone XL protests resulted in 1,253 arrests in front of the White House. I was one of those arrested. The challenge laid at the doorstep of President Obama was for him to take a stand and say no to this project. Public pressure worked, and President Obama did say no; or at least he said "stop"—for now. In 2011, the same year as these public protests, *National Geographic* ran a feature story titled "Anthropocene: Enter the Age of Man," and a few months later, the world surpassed an incredible milestone: 7 billion people. It seemed for a moment that people sat back, thought, and pondered what our huge human footprint on our planet actually meant.

Yet when we look back now, we know that President Obama saying no to Keystone XL at that point and in the manner it was done, was simply kicking the can down the road. During the Obama presidency, the US began the largest and most expansive period of oil and gas exploration and development in US history. Fracking, born under Bush and Cheney, expanded and spread like wildfire under President Obama. Fracking's deadly infrastructure tentacles, including interstate pipelines, gathering lines, compressor stations, power plants and metering stations, began in earnest to grasp, grab, and strangle neighborhoods and towns all across the country. The *all-of-the-above* energy policy pursued by the Obama administration under the guise of energy independence solidified the strength and power of the oil and gas industry—bringing us that much closer to the edge of the climate cliff.

Significant pressure from oil and gas companies was exerted on the Obama administration, and it continues to be exerted in Washington and in state houses across the country today. Many millionaires have been created as Wall Street pushed fracking and its infrastructure forward; money continues to be made. Some of the biggest banks in the world continue to invest heavily in fossil fuel exploration and development and fossil fuel infrastructure.[11] This has to stop. At the same time, fossil fuel companies and their lobbyists continue to push back on the growing scientific evidence about the damages that fracking, specifically, and fossil fuels, more broadly, are causing to our water, our air, and our health, and their role in accelerating our climate emergency.

Harnessing the energy from the sun, the waves, and the wind—in a way that is scalable and renewable—is within our grasp. But it will require decades of hard work to consciously and intentionally make this transition. Success will be defined by bringing along everyone; we can't succeed if winners are picked and losers are cast aside. If we use our brains and well thought-through policies, if we understand truly what is needed—success is eminently doable. Mother Nature already provides renewable energy—cost and pollution free.

Before and during the Obama era, the science generally (and, more specifically, the building body of research on the causes and the seriousness of our climate crisis) was accepted and acknowledged by the vast majority of governments around the world, even if most weren't acting with the urgency required. Our lives seem long, and the rapid changes to our planet still seem to be unfolding in slow motion. Yet our ability to change the future is quickly moving beyond our control.

A new and unanticipated roadblock to swift action on the climate emergency is the discounting of proven science by the Executive Office of the US presidency. In addition, numerous departments of the US national government, including the Environmental Protection Agency and the Departments of Interior, State, and Energy under the Trump presidency, regularly counter and sow doubt and confusion about accepted climate science. A glaring indication of how science is being threatened was the issuance of a joint statement in June 2019,[12] by the US National Academies Presidents of Sciences, Medicine, and Engineering:

> We are speaking out to support the cumulative scientific evidence for climate change and the scientists who continue to advance our understanding. Scientists have known for some time, from multiple lines of evidence, that humans are changing Earth's climate, primarily through greenhouse gas emissions. The evidence on the impacts of climate change is also clear and growing. The atmosphere and the Earth's poles are warming, the magnitude and frequency of certain extreme events are increasing, and sea level is rising along our coasts.

These straightforward facts have been acknowledged and accepted for decades. Yet, the fact that Presidents of the United States National Academies of Science, Medicine, and Engineering felt concerned enough about the clouding of truth to issue this statement in 2019 speaks volumes about how the reality of our climate crisis has been politicized and is now being viewed through a partisan lens. It would seem impossible to be any more direct, clear, or unambiguous on the causes and impacts of climate change. We must break this cycle of climate denial, take back reality, and demand that everyone—regardless of political party—act on our climate crisis with the urgency it demands.

Growth

Our modern society is built on the model of upward growth—quarterly earnings, Gross Domestic Product (GDP) assessments, and a constant drumbeat that *if we aren't growing, we aren't succeeding.* This model, like fossil fuels and dinosaurs, had its place in our history, but it doesn't fit any more. Today, this model has contributed to massive income inequality, destruction and mismanagement of our natural resources, and the perpetuation of our climate emergency.

Wisdom often comes with the benefit of age, experience, and hindsight. Our current model for a healthy and productive society is built on an equation that factors in only a nation's public consumption expenditures; gross private domestic investment; government consumption, expenditures, and investment; and the net export of goods and services. This GDP measurement relies on constant growth to signify improvement. Yet, as world population rapidly expands (with predictions of more than 9 billion people on the planet by 2050), we can no longer continue to use and abuse our planet's resources without significant consequences. Our current model is outdated and not sustainable.

English economist Kate Raworth has asked us to consider her theory of *doughnut economics*, whereby the doughnut is "based on the powerful framework of planetary boundaries but adds to it the demands of social justice—and so brings social and environmental concerns together in one single image and approach. It also sets a vision for an equitable and sustainable future, but is silent on the possible pathways for getting there, and so the doughnut acts as a convening space for debating alternative pathways forward."[13]

Raworth compares unending GDP growth to metastatic cancer. Cancer thrives on the constant growth and multiplying of cells that ultimately destroy their host. She reminds us that there is a growing school of economic thought that is studying the concept and demand for constant growth based on resource extractive industries that treat our planet as a 24/7 dollar store. This research is coalescing around the conclusion that the concept of constant growth is killing our planet and, by extension, us. Therefore, we need new, holistic, and broadly accepted sustainable developments models to follow, grow, and emulate across the globe.

An interesting model to keep an eye on is the 2019 New Zealand *well-being* budget. The budget requires that all new spending go toward five specific well-being goals:

- bolstering mental health
- reducing child poverty
- supporting indigenous peoples
- moving to a low-carbon-emission economy
- flourishing in a digital age

To measure progress toward these goals, New Zealand will use 61 indicators tracking everything from loneliness to trust in government institutions, alongside more traditional issues like water quality. These ideas build on the 1972 Club of Rome report *Limits to Growth*, which measured the use and abuse of natural resources and the environment. The report brought into focus the dangerous path that the world was traveling on, a path built on mass consumption and unending growth. The report made the prediction that our business-as-usual track would result in overshoot and global economic collapse by 2100. Today, almost 50 years later, many of the report's predictions ring true.

An early United Nations metric that set the stage for new measurements for growth was the 1990 United Nations Human Development report. It was revolutionary for its time, showcasing ways for "expanding the richness of human life, rather than simply the richness of the economy in which human beings live." It was and is an approach that focuses on people and their opportunities and choice.

Following along from these earlier metrics, in 2009, the Stockholm Resilience Centre, in cooperation with a team of international scientists, launched the concept of nine planetary boundaries within which managed growth could happen. The premise is that as long as humans remain within these boundaries, life on Earth will continue to function. However, passing through even a few of these nine boundaries could put the planet out of balance, and, in the worst case, create an unsustainable home for human beings.[14]

Other examples of thinking outside the GDP-growth-model include

- Bhutan's Gross National Happiness (GNH) Index[15]
- The 2011 OECD member state well-being report: "How's Life"[16]
- The United Nations annual "World Happiness Report"

One of the newest models for sustainable growth that is attracting a lot of international attention is the United Nations Sustainable Development Goals. Adopted in 2015, these 17 goals "provide a shared blueprint for peace and prosperity for people and the planet, now and into the future. They recognize that ending poverty and other deprivations must go hand-in-hand with strategies that improve health and education, reduce inequality, and spur economic growth—all while tackling climate change and working to preserve our oceans and forests."

Throughout human history, human advancement has been brought forward through change; the constant in our lives and our history is that things don't stay the same. For many people the world over, the idea of constant growth and change equates to increased wealth. But we are now hearing from scientists across a wide range of disciplines, Indigenous peoples from around the world, and religious leaders and ethicists, all of whom are reminding us of a similar prediction: money on a dead planet is just paper and metal. Money will not save or help anyone, and, unless we change our current predominate GDP model of growth, our planet will no longer be able to sustain us.

Corporate Control

We can't minimize the urgency with which we must act. Clearly, we must do everything and do it with extreme haste. Yet, emphasis on personal actions alone is at best naïve, and, to me, it seems to have been embedded in our lives—with sinister intent. The weight of the world is being placed unfairly and squarely on our shoulders, often by the very companies and government officials that could and should be taking on these massive problems directly and urgently themselves. Instead, they are perversely placing these heavy loads on us. They call for actions by individuals that are too much to bear and impossible to achieve given the immensity and urgency of the crisis we face.

In 1953, the Keep America Beautiful (KAB) campaign was launched in the United States to encourage people to *promote a national cleanliness ethic*, based largely on taking personal responsibility for waste. Reduce, Reuse and Recycle become the mantra—one that is still repeated widely and loudly today. Many of the KAB campaign's original corporate founders were from the beverage and packaged goods industries: Philip Morris, PepsiCo, Anheuser-Busch, and Coca Cola. Today, the board of directors of KAB includes many of these same companies, including executives from Keurig, Dr. Pepper, McDonalds, Coca-Cola, Dow, Mars Wrigley, and the American Chemistry Council.

Think about this for a moment.

Beverage and packaging companies began to expand globally just as plastics entered the scene, causing costs to go down as manufacturers transitioned from recyclable and reusable glass and aluminum containers to cheap, single-use plastic containers. Marketing teams arrived at a creative and powerful way to keep costs low—pass on the disposal and recycling of these items to the consumer. Clearly a genius idea, it put the onus on the consumer and made them think it's not only their responsibility but their patriotic and ethical duty to keep the country tidy and clean.

The cost of recovering, reusing, and recycling single-use plastic isn't ordinarily factored into their price. This is how single-use items are kept cheap and accessible. For more than 60 years, the ad campaigns encouraging us to "Keep America Beautiful" have been widely successful. Purchaser beware. You aren't just getting a minute of hydration from that plastic bottle of water; that package in your hand is now *your* responsibility. Find a place to recycle it, or throw that container in the garbage and feel guilty—it's up to you. As it now turns out, facilities to recycle single-use plastics are few and becoming fewer. In 2018, many of our normal recycling venues—plants that had been moved not only out of sight but across the oceans to China, Vietnam, and other Asian countries—stopped taking recyclables from most developed countries. These Asian countries put up *closed* signs and suggested we clean up our own mess. As Canada, the United States, and many European countries scramble to figure out new outlets and economical ways to recycle, Asian and African countries are saying with increasing authority: *stop sending us your garbage!*

In a similar way and with similar ad campaigns, we have been convinced by fossil fuel companies that it's okay if they produce products that cause immense global pollution impacting our health and our planet's health. We have accepted, consciously or subconsciously, that the onus to reduce, reuse, and clean up the messes their products cause has been transferred to us.

The use and burning of coal, oil, and gas creates greenhouse gases and pollutes air and water; this is a fact. For a variety of reasons, however, the responsibility for addressing the damages these gases generate, including the health impacts of hospital visits and the treatment of chronic diseases like asthma, certain forms of cancer and even epigenetic changes (changes to our DNA), are not apportioned; they are just accepted. CO_2 and methane are two of the most damaging greenhouse gases to our atmosphere. Yet, we have allowed our atmosphere to be treated as if it were a giant open sewer, with no collection fees attached. New studies are linking exposures to various types of pollution and stresses, including air pollution and extreme weather, to in vitro damages. Even the place where we consider our children safest is no longer so safe.[17]

Human-created gases have been accumulating in large quantities in our atmosphere, many since the dawn of the Industrial Revolution, and many will linger there for decades and even centuries. For the most part they are odorless and colorless, and their impacts seem dispersed and diffused—out of sight, out of mind. Perhaps, if we colored them pink or purple...?

Think for a moment, about how the emissions emitted from a car ride you took with your grandpa and grandma when you were little are still trapped in our atmosphere, heating up our planet. Like most of us, you probably never gave a second thought to the fact that these gases were being released in the first place, or if you did, it didn't seem like something you had any control over, anyway.

If you have traveled on a river or the ocean for any length of time, you may be familiar with the mantra: *The solution to pollution is dilution*. But at what point does the river, ocean, or our atmosphere reach its saturation point, where dilution can no longer solve or hide a growing and dangerous problem? This point has been reached by our atmosphere, and, as a result, our planet is decidedly out of balance. Oil and gas companies have been working covertly, behind the scenes, and directly in front of you, to keep these facts hidden. Campaigns that put that guilt and responsibility on your shoulders are everywhere: Eat less meat! Use less electricity! Drive less! Fly less! The guilt and burden of greenhouse gas accumulation has become yours to carry.

We must change this.

Putting the onus on each one of us, while hiding the fact that our planet is seriously out of balance and close to spinning out of control, is not just the message of oil and gas companies. Many individuals, corporations, and government officials (for a variety of reasons and beliefs) are working against the inevitable renewable energy transition. They are working overtime to keep things the same—maintaining the status quo—and slowing progress on climate action for as long as they can. Along the lines of the Keep America Beautiful campaign, the concept of shaming or blaming individuals for their personal actions is alive and well.

This blame and shame sometimes comes from the people you might least expect it from, including those deeply involved in the climate movement. Those actively fighting the climate crisis subtly and sometimes overtly place a kind of purity test on one another—pointing fingers at their peers for flying in airplanes, for driving cars, for eating meat, and by doing so, for being less committed than they *should or could be*.

I keep this uncredited post I read some years ago on a friend's Facebook page handy:

> I always have the same internal response, and I usually keep the snarkier parts to myself. But, it goes something like this: "How about I stop breathing air as long as it is dirty? Or maybe I should refuse to go to school or vote or obey traffic lights or pay the electric bill because all of those things are wrapped up in systems of oppression that hurt other people even when they help me."
>
> Here's the thing I often actually say: "I am stuck in this system just as deeply as anybody else, and I could spend all my own time and resources (at whatever levels I have those things) making myself as pure as the driven snow—personally free from the taint of reliance on fossil fuels on almost every part of my life." And by the time I have finished such work,

> I will have wasted months or years of my life trying to make myself above reproach, and everybody else who was stuck in those systems will still be stuck. I will not have developed any power to shift those systems in a meaningful way for a meaningful number of people.
>
> Call me a hypocrite if you want, but I am interested in dismantling the business as usual fossil fuel economy that permeates our lives and tells us that it is useless to imagine another way. I am not interested in achieving my own near perfection so I can look down my nose at you for not coming with me or not having the resources to do the same. This isn't about opting out. This is about building an equitable and livable future.
>
> So, I will do my best to consume less. And you can keep calling me a hypocrite if it makes you feel better. I am about culture-shifting because there is a lot we need to change in the name of justice.

This constantly reminds me that—while doing our best to walk the walk, setting examples for ourselves and others as we reduce our personal footprint—no one is or can be perfect. We need system change at a grand scale. Who has the right or authority to be the judge defining what is the perfect example of individual bests? Having a zero ecological footprint is at odds with being a part of society and living amongst and with community—let's work to make this the norm but recognize, too, that it isn't currently possible in 99% of towns and cities. The problems we face are now so large we desperately need systemic change, and we need this urgently. Lowering our own ecological footprint is clearly something each of us should work toward; however, working for perfection is another thing entirely.

From where I sit, my message on this is: definitely continue to teach your children to recycle, to tread lightly on our planet, and to measure and reduce their personal and your family footprint; this is, and should be, part of our parental responsibilities. However, because society didn't act soon enough, our climate crisis is already reaching tipping points and points of no return. Bringing our planet back to a state of balance will require collective and societal actions on a global scale. We need policies and laws that will impact us all and move all of us in these directions. We also need laws that require corporations to clean up their own messes rather than pass on their messes to us. At our current individual pace, we will not get to where we need to be in time. But this doesn't mean we should throw up our hands and do nothing. We must and can lead by example. Individual responsibility—changing our habits and reducing our personal footprint—is important. We can all be part of the solution.

Most recently, we've been talking a lot about how the planet has limited resources and how it is the responsibility of us all to use these resources wisely. This can be difficult in practice because it goes against many of the social norms in our community. It means not buying all of the toys that he wants, not distributing goodie bags at the end of birthday parties, and not purchasing plastic bottles in stores. Sometimes I feel like a Scrooge, but I want him to grow up with a realistic picture of what I believe the world will look like in the future. A world where resources are valued and protected, where we will buy less, buy used, and where sharing and repairing are common practices. I don't want him to look back at this age as the time when he had everything and then one day it all disappeared. Instead, I want him to learn to appreciate what he has, but also value time, friendships, family, public spaces, and nature. I want him to learn that life can be meaningful and rich without so many material items.

—Jill Kubit

These nuanced concepts are ones that for many of our children will be difficult and perhaps impossible to fully explain. For many of us, they are hard to come to grips with. So do continue to teach and model by setting positive examples. Talk about climate change as you show your children you are taking steps to reduce your family's footprint; and work actively on explaining, activating and demanding transformational, societal change.

Justice

A just path forward, climate, and environmental justice—these are words used and heard often. What do they actually mean? According to the Climate Justice Alliance, a just transition is "a vision-led, unifying and place-based set of principles, processes and practices that build economic and political power to shift from an extractive economy to a regenerative economy. This means approaching production and consumption cycles holistically and waste free. The transition itself must be just and equitable; redressing past harms and creating new relationships of power for the future through reparations. If the process of transition is not just, the outcome will never be. A just transition describes both where we are going and how we get there."[18]

Climate justice connects environmental justice—including the right to clean air, clean water, a healthy environment and food security—to human rights. It recognizes institutionalized and historical injustices that perpetuate and exacerbate poverty through global—national, regional, and local—actions that have local implications. Our climate emergency is not color blind; nor is it an equal opportunity crisis.

Just as Black Lives Matter is not the same as All Lives Matter, the impacts of the climate crisis, and the need to ensure a just transition, remain critical to our ability to successfully address the crisis at hand. Yes, climate change is happening all around us and impacting us all. Yet, those who can afford to protect themselves from the direct impacts of our growing crisis are still often able to and for the most part, do so. Those who can't, are already—and, often, deeply—reeling and suffering the consequences.

Why is it that in the United States of America, African Americans are almost *three times more likely* to die from asthma-related causes than white Americans?[19] Could there be a correlation between the fact that the vast majority of air-polluting infrastructure, such as power plants, cement and chemical processing plants, and incinerators, are more likely to be sited in low-income neighborhoods that have historically not been able to gather the community support or the political attention to stop their construction? In fact, a 2016 study by the University of Michigan found that *hazardous waste sites are often built in neighborhoods where whites have already been moving out, and poor minority residents have been moving in, for a decade or two before the project arrived*. This follows on the seminal 2007 report, "Toxic Waste and Race at Twenty"[20] that found that more than half of all the people in the US living within two miles of a hazardous waste site, were people of color.

So, as we talk about a range of climate solutions, about jobs in the new green economy, and about infrastructure and public transportation, who gets those jobs and where that new infrastructure is to be built needs careful consideration, thought, and planning. As the economic gulf between rich and poor is exacerbated, how we work to close this gap as we build a new and hopeful future—one that isn't built on resource-extractive, polluting industries—will be key to helping everyone participate and benefit in this new future.

How, too, do our moral and ethical responsibilities stack up with the practical implications of the growing and destructive impacts multiplying *over there*? Where actually is *over there*? We must open our eyes to the reality that over there is just as likely to be across town or down the block than it is to be in a country across the ocean. A big question to ponder is, Are we willing and able to reconcile the importance of addressing climate justice along each step of our journey?

The impartiality of the climate crisis will only grow stronger, yet the fact remains that many black, brown, and low-income communities are the ones that are the least resilient, those least responsible for our climate crisis, and those hit first and worst. These communities cannot easily *pick up the pieces*. Maybe they can't do it at all. Many families in Puerto Rico and in parts of Texas and in Florida are still reeling from Hurricane Maria's 2017 destructive forces;

I have rearranged my life to center climate action with a social justice lens, and I am now a community-supported organizer. This work has taken a significant financial toll on my family as existing institutions have proven ill-suited to address the problem at scale, or acknowledge the root causes of it. In the past it was easy to include my wife and son Trevor in climate actions, but now it is more difficult with two additional children. Also, the work that I believe we need to be doing now involves less protesting, and more direct actions and community building. As a result, I haven't been able to include them as much in the work. I am not interested in showing them that I've 'got this,' as I don't think any of us can really say that we do. But I would concede that it is important to be able to say that we did everything we could when we knew.

—Derek Hoshiko, organizer, For the People, Clinton, Washington

recovery isn't equal or just. When a tragic fire struck Notre Dame Cathedral in 2019, billions of dollars were committed to help rebuild it within days of the tragedy. Yet, when Mozambique was struck with devastating cyclones, leaving hundreds of thousands homeless and without food or water, also in 2019, many around the world watched, and most looked away. Resources commensurate with the disaster at hand were not given or even promised. Why?

This reality plays out across the US on a regular basis. Louisiana, Puerto Rico, Florida, Nebraska, North Carolina, and other parts of the US south and Midwest continue to be hit harder and harder with each passing climate-exacerbated disaster. Yet we continue to *hide* the impacts of climate disasters in North America, even as people suffer daily and recover more slowly, if at all. Conveniently, media attention comes and goes quickly. Building the resiliency needed to protect from our coming storms also isn't being done in a just manner, even as it is beginning to be put in place in wealthier communities and cities.

All over the African continent, in the Middle East and in Central and South America, if we open our eyes, we can see that climate-induced famines, droughts, and wars have created situations in which people are regularly—and with increasing frequency—displaced from their homes. Daily, people are starving; they are living in war-torn areas, often without access to clean water, health care or safe havens. We must remember and remind our children that focusing our lens through justice demands that there be no sacrifice.

Each of us deserves to be treated with dignity, respect, and care. This isn't, however, the world we live in. Our world is unjust, unfair and unequal—and it is becoming more so. Being born into a certain zip code, town or country shouldn't define your chance at an education or your job success; nor should it be a marker of your ability to be able to be resilient in the face of our climate crisis. However, where you are born more often than not does define these outcomes, in disproportionate, unfair, and unjust ways.

This small section of this one chapter cannot hope to, nor does it even attempt to, do justice to a discussion on climate justice. Hopefully, it does raise questions for you. Do seek out more information and more resources so you can have thoughtful conversations with your children about how climate justice is an integral and critical part of all successful climate solutions. Check out the resource section at the end of the book for places to start.

As John Muir said back in 1869,

> When we try to pick out anything by itself we find that it is bound fast—by a thousand invisible cords that cannot be broken—to everything in the universe.

We are finding ways to cut these invisible cords that the planet has carefully crafted over millennium. As quickly as we have appeared on planet Earth, we could disappear too. The planet will heal. The question is, Will it do so with or without us? Depending in large part on what we do and how we react today, we can still have a hand in creating and shaping our future. It isn't predetermined. We don't have to go down the devastating rabbit hole in front of us. But we need to begin now.

3

Coming to Grips with Climate Change as a Parent

From Grief to Hope to Resolve to Action

For many new parents, the birth of their first child is what brings the urgency of the climate crisis immediately home and lights the spark that then guides their future personal and collective actions. For many others—me included—when my children were first born I didn't think too much about the climate crisis. Even though I was working on it from a policy side, it still seemed to me to be at least *seven generations away*.

> I think many, many people continue to feel this way today.

Coming to Terms with Reality

Upon becoming a parent, my immediate concerns involved making sure my babies' world—our New York City apartment and everything that touched it or touched my kids—was as safe and secure as I could make it. This, to me, seemed something I could directly control. I researched additives in foods, organic vs natural, cotton vs synthetics, plastic baby bottles vs glass, side sleeping vs back sleeping, disposable diapers (which kind, not if I should or shouldn't), vaccines, sunscreens, bug spray (or not), bedding, coffee table protective corners, high chairs, and the list went on and on. Clearly, many of these were and still are valid health and safety considerations for parents the world over, and certainly they are not to be taken lightly or ignored.

However, in spite of our attempts to keep our children safe, in the United States and in many parts of the world, outdated and ineffective chemical regulations have turned our children's bodies into active experiment zones. We know that in the US over 50,000 chemicals in use prior to the 1976 Toxic Substances Control Act (TSCA) remain in use today and upward of 80,000 chemicals impact our lives daily—all untested and not proven safe for humans. The 2016 Frank R. Lautenberg Chemical Safety for the 21st Century Act was introduced with fanfare and high hopes to update, correct, and strengthen the outdated and ineffective TSCA act.[1] Unfortunately, the effectiveness of the Lautenberg Chemical Act remains largely unproven and unseen. The speed of its implementation is exceedingly slow, with few results and limited or no success in removing known harmful chemicals that are already on the market.

Connecting the dots to our climate crisis, we know that many of the most hazardous chemicals in our lives today are derived from fossil fuels. More complicated but equally relevant is the fact that our children—in fact, all of us—are part of a grand experiment. We are at the mercy of companies that aren't really sure how harmful their products are, yet they are willing to bargain with our children's health while they, and we, wait and see. This same attitude and lack of adherence to the Precautionary Principle[2] is ingrained in the mining, production, distribution, and transportation of fossil fuels, as well as with production and use of Big Agriculture products. These experiments involve not only our children's health but that of our planet.

By taking climate action that builds solutions, we are showing our children that even if it's hard work, there are ways to move forward. All is not lost. We cannot—nor will we—give up. Sharing opportunities, stories of success, creative and positive actions—all will build active hope.

We choose to go to the moon in this decade
and do the other things, not because they are easy,
but because they are hard; because that goal will serve to
organize and measure the best of our energies and skills,
because that challenge is one that we are willing to accept,
one we are unwilling to postpone, and one we intend to win.

—JOHN F. KENNEDY, 35th president of the United States

Our children are literally in mortal danger, and yet there are many parents whose mama and papa bear reflexes haven't even began to kick in. What's holding them back? How is this possible? Clearly the stakes have never been higher, yet humanity is trying to rig the game against itself. By refusing to acknowledge what's right in front of us, we risk losing everything—happiness, health, home, and a stable and secure future. To win, we must teach our children and remind ourselves of three simple life lessons:

1. **Tell the truth.** There is no longer any room for denial around the climate crisis. We humans are causing our climate to change—period. The sirens are deafening. The evidence is overwhelming. End of story.
2. **Actions speak louder than words.** We must acknowledge and recognize that there is no bridge to a renewable future that is built with a continued reliance on any form of fossil fuel. We need to step bravely into the abyss, trust in science and the evidence, and make the leap—with individual and collective action. This will put people to work and begin to heal our planet.
3. **Don't be afraid.** We must look "truth" squarely in the eye and not be afraid. Scientists are telling us and our planet is showing us: we need to act. Together with our children, friends, family, and all humanity, we need to move quickly and boldly forward to reclaim a livable future.

> I am hopeful that the odds are changing, ever so slightly, in humanity's favor. More and more caring and thoughtful people are seeing the emperor in the full light of day, standing up to him and demanding that others open their eyes and see him clearly too. Together we can and must change the collision course we are on; there is no other option.

I wrote this for a publication called *Global Chorus*[3] when my children were just entering their teens. Except for a few updates, I could have written this today.

Time flies.

Unfortunately, our burnout, sadness, and grief too often creep in as time passes and as our planet grows sicker. It's so easy to get despondent, melancholy, depressed, and overwhelmed. How do you handle this? I make a conscious effort to remember to show kindness, love, and gratitude to others working to protect and preserve our future and our present. Whatever we do, and however we do it, it is a forward motion. Treating each other with grace and appreciation is critically important. These are lessons parents have taught their children throughout history; we must model them today for our children, and not allow them to be lost.

Yet today, all around the world, we are witnessing how easily acknowledging and sharing feelings of appreciation and thanks are being thrown aside—sometimes carelessly, sometimes in undue haste because of the crisis at hand. Sadly too, sometimes this happens and is on display with forceful and mean-spirited purpose, without any obvious reasons. The examples for our children, of powerful people treating each other with disdain and disrespect are on view regularly. So, in these divisive political times, as the wealth gap between rich and poor widens, and as we open our eyes to the existential threat that requires us all to come together, remember to lead through examples of kindness, love, and gratitude.

What's tough for me about talking to my children about the climate emergency is that I hate to pierce the bubble of childhood innocence. We want our kids to feel safe and secure and comfortable. Of course, it is the privilege of my family's economic security, nationality, and skin color that shelters us [from] experiencing most of the impacts of climate change right now. But climate conversations now fall in the same category as the honest reckoning parents owe their children—and our society—about racism and slavery, sexual abuse, misogyny, and homophobia—or guns and cyber bullying. The world they must navigate as they grow is magical and full of wonderful people and places, but it is not always pretty. Opening their eyes to all that's ugly and dangerous makes them better people—hopefully equipped to be resilient, to be emotionally awake, to be connected to people and their world, and to make a difference for good.

—Anna Fahey

What is that driving force that, once you are fully aware, won't let you look away? Is it grief, hope, or resolve? And does it really even matter which of these strong feelings are driving your actions? I think it is all three. These feelings may be apparent and ever present in some individuals and at other times each feeling is expressed on its own, in its own time and space. Validate and acknowledge your feelings so you can move to the next stage: action. This means allowing yourself to feel the immense grief that comes with understanding what is happening right now to us and our children, and to acknowledge deep down, that we know in all likelihood what lies ahead.

> For some, accepting the reality of our climate emergency may mean arriving at a decision to not have children at all.

Most of you reading this book have children of your own. Or maybe you are an uncle, an aunt, a teacher, or a dear friend to children who are a large part of your life. If you have chosen to bring children into the world, you may or may not have a clear understanding that their childhood and their future, because of our climate crisis, will be more challenging and in many, many ways much different from your own childhood experiences. More frequent extreme weather, increasing turmoil brought on by climate stresses (including refugee crises), food shortages, air and water pollution, sea level rise, and more will all become even more commonplace then they already are, as conditions worsen.

Given our reality, there are more and more people who, upon clearly understanding the scientific forecasts and our unfolding climate crisis, have decided not to bring their own children into this world. This is understandable for those who feel that our climate crisis presents an unfair burden and potentially a chaotic or perhaps catastrophic future for those alive today and for those yet to be born.

Giving those who have made or are weighing these decisions the respect due is something I recommend we all aspire to do. It truly takes a village to raise a child, and being a parent isn't a prerequisite. Recognizing that there is no one way—no linear path or single *correct* way—is important. Giving space to mourn what we know we have lost, accept what is at stake, and create room for conversations about what family looks like in the 21st century is critical for our growth and our development and for incubating and developing thoughts and ideas to share with our children.

Grief

Most of us know intimately how easy it is to be swallowed up and paralyzed by overwhelming sadness. I experienced this when my father was dying from cancer. Unable to change the inevitable outcome, it was hard to find joy in anything. My escape was physical distance, which allowed me to shut out the truth and the day-to-day realities of his illness.

Out of sight and out of mind rings true with the climate crisis as well. For those of us who live in the United States, Canada, Australia,

Europe, and most developed nations, it often seems like the climate crisis is happening *over there*, not in our own backyard. Accepting and understanding the depth and the urgency of the climate crisis—a crisis of species survival and a palpable existential threat—enhances and perpetuates our feelings of helplessness and grief. Often, it seems best to just close our eyes, shutting out a reality that just doesn't seem possible. Scientists and sociologists often use the phrases *motivated avoidance* and *apocalypse fatigue* to explain this phenomenon. Very often, our brains will just shut down when the subject of climate change is raised. Our climate emergency feels too big and too out of our control, so we literally close our minds to the crisis—shutting off the search for solutions as well.

My anxieties about the future often manifest through thoughts that demand I have answers immediately. Do you feel this way too? Do you demand assurances today so that you can relax just a little bit, secure in the knowledge and understanding that our unsettled world will at some point become more settled and ultimately a livable place for our children? Instead, because of the warning bells from scientists, economists, policy advocates, and activists—we must let go. We must have infinite patience and let the future unfold; the assurances we are looking for are not coming. We may never see the solutions we seek; nor will we ever arrive at certainty about the future to come. Some call this faith in the future—intrinsic hope.

The grief and sadness we feel from understanding the depth of our climate crisis has its own unique name: *solastalgia*. The term was coined by an Australian environmental philosopher and professor of sustainability, Glenn Albrecht.[4] Solastalgia refers to "the homesickness you have when you are still at home." According to Albrecht,

> Solastalgia has its origins in the concepts of "solace" and "desolation." Solace has meanings connected to the alleviation of distress or to the provision of comfort or consolation in the face of distressing events. Desolation has meanings connected to abandonment and loneliness. The suffix -algia

> has connotations of pain or suffering. Hence, solastalgia is a form of "homesickness" like that experienced with traditionally defined nostalgia, except that the victim has not left their home or home environment.[5]

Around the world, psychologists and psychiatrists, in their day-to-day practices and through collaborative work on peer-reviewed papers, are looking for ways to address the growing mental health issues associated with climate change.[6]

It has been my personal experience, over the years as a climate activist and educator, that people have long talked in whispers about the prospect of human extinction—seemingly testing one another. Could this really be a possibility? Could our species' extinction actually be the end run of the crisis we have set in motion? We used to think we shouldn't say such things too loudly, or people would think we were crazy. This is no longer the case. The whispers are becoming shouts, difficult to ignore. Extinction Rebellion and school strikes have exploded on the scene. The Climate Mobilization and its drumbeat to create a World War II-style mobilization to address the climate crisis is being heard, and hundreds of cities around the world, including big cities like New York and London, have passed climate emergency acts.

Species extinction, including ours, is talked about with more directness in many circles. In formal scientific and academic settings, this eventuality is being discussed with openness and without hesitation. Human extinction is a clear and possible outcome of our growing climate crisis if we don't begin immediately to treat climate change like the emergency it is. Yet, for billions of people, and for a wide range of reasons, this reality is still not on their radar. Or if it is on their radar, many still identify the people raising the alarm bells as lunatics.

So, even as we acknowledge and allow ourselves to feel the weight of the sadness that our climate crisis has laid bare, I believe we must consciously validate and make room for active hope. We must find ways to build this, as we nurture the determination and resolve that will move us forward, so that these feelings can find a way to settle into our lives. Each and every feeling is real and must be acknowledged and affirmed for us to be able to take the next steps forward. Accepting the depth of the crisis and acting on this in a commensurate way—without losing our sanity, our hope or our resolve—is how we will keep going.

Five Stages of Grief

In 1969, psychiatrist Elisabeth Kübler-Ross identified the five stages of grief from her observations working with terminally ill patients: *denial*, *anger*, *bargaining*, *depression* and *acceptance*. Once we fully understand the urgency and the depth of the multi-layered problem that is our climate crisis, many of us go through these same strong feelings and stages.

In a nutshell it often goes something like this:

- **Denial:** This can't be true! Scientists are making this up. This will happen so far in the future. My great-great grandchildren will be able to figure it out.
- **Anger:** Why haven't THEY done something?! Who can I blame....?
- **Bargaining:** I'm going to stop eating meat; my next car will be electric; I will stop flying. If all my neighbors do this too, we will solve our climate crisis. I know that there are many people working on solutions. I'm sure THEY have got this under control!
- **Depression:** We are out of time. I see it and feel it. I'm panicking. I can't catch my breath. I can't sleep. I dream of climate disasters. Where can I hide? Where can I go? How will I protect my children?
- **Acceptance:** I understand. I see the impacts directly. I realize how serious this is and that it isn't something that one person or one government will solve. We will be living with our climate crisis for the rest of our lives. We need to get comfortable with this very uncomfortable fact.

Coming to terms with your grief and accepting our new reality are critical steps to take before talking to your children about living with and through the climate crisis. Once you are comfortable with these uncomfortable truths, you can find ways to help your children find their own way to acceptance, and beyond.

The road to acceptance begins with telling the truth. For most of us, our natural instinct—housed in our genes and honed when we were hunters and gatherers—is to protect our children at all costs. We are moved to do this instinctually, both physically and mentally. We search frantically for ways to let our children know *we've got this*, even as we know deep down that we do not—at least not yet. At a certain age, our children are old enough to understand that we are living a climate emergency. This will likely be when they are younger than you think they would (or should) be able to grasp the crisis we face. They will look to you for advice on how to accept this and how to move from fear and despair to hope and action. This age will be different for each child, even those in the same family.

Jim Poyser, a wise counsel and dear friend from my Climate Reality family put it this way:

I work with students of all ages in various contexts, including school visits. Recently, I was speaking to a group of high school students and was waxing, apparently, too positive about our prospects to solve the climate crisis. One of my key youth leaders, a high school student, was in the room and called me out in front of everyone, taking issue with my blithe remarks. Later I asked her what I should say to her peers, and she replied, "Just tell them it doesn't look good right now, but we have no choice but to try." I thought that was pretty good wisdom there from a 16-year-old; if kids really want to know what's happening, you have to be honest with them. But most importantly, you have to give them opportunities to meet the crisis with something meaningful.

—Jim Poyser, Executive Director, Earth Charter Indiana

Wise advice: *You have to give them opportunities to meet the crisis with something meaningful.* I believe that to do this, we must first find ways to do this for ourselves. We know that the compound problem of climate chaos is edging forward faster than we can imagine or ever thought possible. We do still have time. So to move forward, we must begin to visualize the future we want, telling the story of what will be and how we will get there to ourselves and to our children, so we can and will make it happen.

I have discovered through my years working with Climate Mamas and Papas around the country and around the world, from long talks with my Climate Reality family, and through heart-to-heart conversations with dear friends and colleagues from across many of my *lives*, there are several more stages that come after acceptance—that moment when we truly accept the grief that comes with understanding the depth of our climate crisis. That moment is followed by *hope, resolve and action*.

Hope

> *Hope is a renewable option:*
> *If you run out of it at the end of the day,*
> *you get to start over in the morning.*
>
> —Barbara Kingsolver, American author

Hope can't exist in a vacuum; it needs air to breath, just as we do. Each one of us likely defines and practices hope in different ways. We each will derive strength to move forward from a range of perspectives. For some, this will be spiritual or religious. For me, it's active hope. Eco-philosopher Joanna Macy, in her book *Active Hope*,[7] teaches us how to channel and manifest hope. Using hope as a guiding light, she helps us look directly at our unfolding climate emergency and learn how to move forward:

> Active hope is a practice. Like tai chi or gardening, it is something we do rather than have. It is a process we can apply to any situation, and it involves three key steps. First, we take a clear view of reality; second, we identify what we hope for in terms of the direction we'd like to move in or the values we'd like to see expressed; and third, we take steps to move ourselves or our situation in that direction.... The guiding impetus is intention; we choose what we aim to bring about, act for, or express. Rather than weighing our chances and proceeding only when we feel hopeful, we focus on our intention and let it be our guide.

Another form of hope that leads people forward through our climate crisis is intrinsic hope. Author Kate Davies helps explain this concept in her book *Intrinsic Hope: Living Courageously in Troubled Times*.[8] Davies tells us,

> Intrinsic hope is a deep abiding trust in whatever happens and in the human capacity to respond to it positively. It accepts life just as it is and works with it, whether or not it's what we want.

The concept of intrinsic hope explains the capacity to keep drawing from a deep well of hope, preserving the belief and solid trust that hope will remain, and will not run out. Intrinsic hope allows us to continue to take the pummeling body blows that our climate crisis rains upon us, giving us the ability to accept that our human species can and will find ways to respond. "Radical hope," a phrase often credited to Jonathan Lear,[9] has its origins in a study of the Crow Nation.

When the buffalo went away
the hearts of my people fell to the ground
and they could not lift them up again.
After this nothing happened.

—Chief Plenty Coups

Lear asks us to look deeply at how we can and will continue, even when we know that our culture is falling apart and cannot be saved. Radical hope must then take over.

So, while many parents have had their eyes open for a long time, more and more people are opening their eyes. This development alone can and should feed our hope—from whichever perspective we define it. Even in the midst of the urgency we face, allow yourself to rest for a moment; remove the weight of the crisis from your shoulders. You are not alone. Take deep breaths; let go of any guilt that you may feel by letting your mind and your heart wander away from the crisis for a brief time. Give yourself permission to take the time to appreciate something funny or something mundane. Our lives continue; we must live them fully and remind our children to do so, too.

Climate Hope: Becoming a Mother for the First Time

I've been back to work for a few weeks now and have had a bit of time to sit and reflect on all of the emotions that come with being a new mom and to consider how my drive to continue our work here [Climate Reality] and fight this fight are amplified, made more urgent, made, unbelievably, somehow even more important.

I woke up last night not to her shuffling in her crib (thankfully she was sound asleep) but to an all-consuming feeling that I must do everything I can to keep her safe.

This will all be wildly obvious for any of you who have children...but my world has changed. Flipped upside down and back again, and new filters put on my vision.

Life is sweeter, yes—but my sense that the world is in dire need of help is also sharper than ever before.

During my pregnancy, I thought often of how I was going to talk to her about the climate crisis. How would I explain the world that we are bringing her into, and the urgency of fixing these problems that can't wait any longer? At four months, she knows no fear, has no sense of danger.

But I do. For her. For her world.

We struggled to get, and stay, pregnant. I share this because of how common it is but also because it made me question if this is how we would create our little family. I wondered if we even should continue after my miscarriage, and I debated if this was some sign that perhaps this simply wasn't right. We made the deliberate decision to keep trying, and because of that decision, we were blessed with our baby girl, but also that decision meant that we are burdened with the undeniable knowledge that we purposefully brought her into this world and have to do whatever we can to ensure it is a world in which she can thrive.

A Climate Reality Leader at the Los Angeles training, when I was six months pregnant, told me I was so brave for bringing a new life into the world in its current state. Is it bravery? Is it naiveté? Is our biological drive to reproduce behind the times, not recognizing the complications of our world? What reason do we have?

The reason I've landed on is hope.

It is hope that brings us to create new life—hope that we will mitigate the problems that we can, adapt to those that we cannot, and ensure there is a world to be cared for far beyond ourselves and long after we are gone.

One of the priceless benefits of my position is that I get to hear all the time about what's being done, in places all over the world, to solve this. And I get to hear about what you, the open-eyed and brave, the global corps of Climate Reality Leaders, are doing each and every day to protect the future for Lani and every new baby born to this planet. And I tell you, it gives me hope."

—Olena Alec[10]

Seek out groups, individuals, organizations, companies, municipalities, states—everyone and anyone in your circles or in circles around you—who is actively taking action to slow down the climate crisis and identifying and implement solutions. From the actions of others, you will find ways to build your power, your hope, and a way forward. If you look, you will find those people and groups in your community, your city, your family, or your place of work. Join them, take action with them. They will inspire you, support you, and help you as you find ways to create actions and active hope of your own.

Resolve

How do we build and sustain the resolve needed to keep taking climate action, to stay involved and to lead? Resolve is the next step. As science and Mother Earth keep *hitting us over the head*, reminding us that our common home is on fire and the firetrucks and firefighters are nowhere to be seen, there will be many days when you feel like screaming, crying, and hiding, many nights when you can't sleep thinking of what lies ahead. Why won't people wake up?

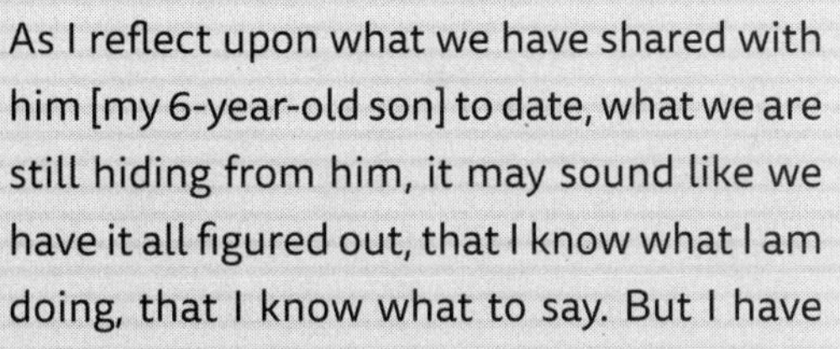

As I reflect upon what we have shared with him [my 6-year-old son] to date, what we are still hiding from him, it may sound like we have it all figured out, that I know what I am doing, that I know what to say. But I have many doubts. Does he know too much? Too little? Are we preparing him for what his future will look like? I honestly don't know. I don't know the answers and don't know if anyone really does—at least not yet. This is definitely new territory. So, in our family, we are just experimenting with what we think our son can handle at various ages and learning from it as we go.

—Jill Kubit

With each passing year, the science is clearer and the messages we are seeing, feeling, and hearing more obvious. Yet the pace of society's response remains very slow—at best. As individuals, as parents, grandparents, uncles, and aunts, this lack of timely response from people around us can be frustrating and often maddening to the very core. We constantly remind ourselves as we build the resolve to move forward, one step or many leaps at a time, that we must keep going. But these forward actions may sometimes feel not only difficult, but next to impossible. So, reminding ourselves to take care of ourselves—physically, mentally, and emotionally— is essential. Regular self "check-ins" are critical for our own success. If we are to be effective, our well-being must be front and center.

My commitment to work to stop global warming came prior to becoming a parent. My motivation comes from a deep moral imperative and spiritual calling. Unlike some folks, while being a parent certainly has put some things into perspective, it hasn't fundamentally shifted my attitude or commitment to the work. I would say, however, that having witnessed more than one climate justice activist either lose their family or have family tensions arise due to their activism, I definitely consciously make time and space for my wife and children whenever possible, i.e., evenings and weekends. Work-life balance, or in this case activism-life balance, has become a priority. It is also necessary self-care to be able to continue showing up for work that is difficult.

—Derek Hoshiko

Even as our world becomes more polarized, more and more people are accepting the reality of what we face, resolved to work on the crisis from whatever perspective and vantage point they can. We don't need everyone, but it helps if we are all traveling in the same direction, even if we take different routes to get there. At the same time, it doesn't mean we should stop talking to those who have different ideas about how to move from grief to hope and resolve. It remains important to weigh ideas, to have difficult discussions, to share concerns and fears with others who understand where you are coming from. This understanding—that other caring, concerned, and activated individuals are working together and apart in a multitude of ways to slow down the unfolding climate crisis—should steady you and build your resolve.

Action

Focus on what you are passionate about. There are so many options out there and so many areas requiring your attention—all of which are intertwined and interconnected. If we each take on those challenges we feel called toward, then we will be successful in slowing down our unfolding climate crisis, turning it into a chronic condition rather than a fatal one. I believe that we all can and must be activists in our own communities—*owning* the word "activist," which will mean different things and manifest in different ways for each of us.

As parents, once the depth of the crisis is clear, our actions must be commensurate with the crisis at hand.

Addressing the climate crisis is often discussed from two sides of the same coin, both of which have a multitude of broad ways to be tackled. One side, the *mitigation side* involves finding ways to amplify and advance solutions that slowdown our crisis by winding down and then drawing down human-created greenhouse gases—ultimately stopping their production and proliferation. This side includes actively working to stop new fossil fuel infrastructure like pipelines; compressor stations; or gas, oil, or coal-fired power plants or working on solutions that draw down greenhouse gases broadly from our atmosphere.

The 2017 book *Drawdown: The Most Comprehensive Plan Ever Proposed to Reverse Global Warming*[11] was a welcome breath of fresh air for those in the climate community, and it remains an active replenisher of climate hope for the world today. Edited by Paul Hawken and compiled with input and consultation from a coalition that included more than 200 researchers, business leaders, scholars, and change-makers from across the globe, the book has been widely fact checked, reviewed, and validated. The solutions presented are viable, scalable, concrete ways to pull down greenhouse gases from the atmosphere and at the same time offer alternatives that will allow us to stop putting greenhouse gases in the atmosphere in the first place. Instead of just focusing on carbon dioxide, *Drawdown* presents ways to reduce and eliminate all human-caused greenhouse gases from the atmosphere. Some of the solutions may surprise you. Shorter-lived gases, like those associated with refrigerants, are significantly more potent in the short term than CO_2. As a result, refrigerant management is listed as the top solution; if we don't manage it, it could be vehicle for greenhouse gases that push us off the *climate cliff*.

Drawdown's top ten solutions are as follows:

1. Refrigerant Management
2. Onshore Wind Turbines
3. Reduce Food Waste
4. Plant-Rich Diet
5. Tropical Forests—and Use
6. Educating Girls
7. Family Planning (when 6 & 7 are combined, they become the number one solution)
8. Solar Farms
9. Silvopasture—integrating forests and pastures
10. Rooftop Solar

The flip side on the climate crisis coin from mitigation and drawdown is the *resiliency and adaptation* side. So instead of putting the focus on controlling and reducing emissions, resiliency and adaptation involve strengthening and hardening our cities, communities, and our families so that we can both withstand and provide safe haven from our escalating crisis. Some of the actions on this side of the coin include enacting and then implementing climate resiliency laws and policies; creating new inventions and ideas and then building on them; transitioning toward a circular economy; planned and managed retreat for those communities in the path of sea level rise; regenerative agriculture; ensuring access to water and food for all; and overtly creating and implementing a just transition. Both sides to the coin will take many forms—as wide and as broad as one can envision, and then farther still. There are so many ways to take action!

Trusted Messengers

Climate communications research reminds us that trusted messengers—those people we seek out and rely on for the truth—are often those closest to us. These messengers will be different people for different individuals and different audiences. Often, trusted messengers are our friends, neighbors, and family members. Our children, as our trusted messengers, are heard more directly, in a far deeper and clearer way. Young children too, have a way of speaking the truth that is obvious, without built-in biases, and in a way that raises no doubt about what they mean. Our children—through their understanding, their direct and clear voices, and their demands to their parents for action—may ultimately be the ones who break through the logjam we currently face and change minds that have so far remained closed.

We had come across this idea that kids
are capable of influencing their parents.
And when we say influence,
we really mean just teaching them.[12]

—Danielle Lawson, social scientist North Carolina State University

---

For the first time since 2007, when I jumped in with both feet to work on climate action, I feel a palpable difference. Active hope is growing, built on a solid foundation. People all over the world are seeing and experiencing the realities of our climate crisis. They are also beginning to imagine what could be beyond the immediacy of the crisis, accepting the chronic condition but not the finality of a death sentence. Children's voices are demanding that we not only listen but that we *hear* them. Our children are growing up living climate change; many are angry, frightened, and hurting. As parents, it is impossible to look away when our children are crying out. Politicians, business leaders, clergy, and a broad spectrum of people across our diverse communities are not only seeing our children but also hearing them. Interestingly, our children's voices may be the tipping point for powerful fossil fuel companies to join the transition that requires unwinding our intricately connected fossil-fuel–based energy system and infrastructure. Or, conversely, our children's voices may be the tipping point that oversees the demise of these companies as they become less relevant as we move forward with a renewable world order. In a 2019 article in *The Guardian*[13] Mohammed Barkindo, the Secretary General of the Organization of the Petroleum Exporting Countries (OPEC) stated, "There is a growing mass mobilization of world opinion against oil which is beginning to dictate politics and corporate decisions, including investment in the industry." He went on to say, "Pressure was also being felt within the families of OPEC officials because their own children are asking us about their future... they see their peers on the streets campaigning against the industry."

According to the United Nations half of the 7.6 billion people alive in the world in 2019 were under 25. However, this doesn't mean those 25 and older are off the hook! Action begins by setting your own personal intentions. Once you can imagine the future you want, you can then help your children find ways to imagine the future they want, too.

4

Leading by Example

Now that you are clearer on what's at stake and why we are facing this climate emergency, I hope that you are beginning to formulate ideas—personal to you—that will help explain to your kids, in your own words, what's happening. If you are already involved in the climate movement you can remind them, or perhaps tell them for the first time, what you are already doing. If you are just beginning to come to terms with the crisis, you can let your kids know you are learning everything you can. And, if you find yourself somewhere in between these two ends of our climate action continuum, let your children know you are researching ways to take action (one way is by sharing what you are reading in this book) and that if you aren't already, you will be soon putting these ideas into practice. Assure them that there are many great examples of families, communities, and individuals young and old working on this full time. Many of your children—like many of us—are angry, confused, and frightened. Reassuring your children, when deep down you often feel a hollow in the pit in your stomach, causing you to wonder if indeed you can keep your promises to them, may seem impossible. So, take a deep breath and do as you tell your children to do: take it one step at a time.

At all times, be honest and provide the facts to the best of your knowledge. Do try your best, when talking and acting on climate, to be in that moment with your child. Just as you sometimes may be a distracted driver, being a distracted climate parent is a common thing in the 21st century. *Let me just finish this e-mail, post, tweet, or text—it's important, it's about taking climate action*. I am sure I am not the only one to say these words to her child. Be conscious of these distractions, and make every effort to be in the moment as best as you can.

We know our planet is currently getting sicker, and our future is indeed precarious; but remind your children—and yourself—that many people do understand this and are working every day to develop solutions and to slow down the changes we have put in motion.

We're pretty open about it with our kids and don't try to sugarcoat it. My son, who's really into the natural world and understands climate change quite well, would likely call me on it if I downplayed it. Probably the most important things I tell them is that there is hope, there are a lot of really exciting developments lately in the climate movement, and that they can do something about it.... They [my children] know that the situation is not great, and there's a lot of work to be done. But, I think, them seeing me work on it, seeing positive change locally, and hearing about some of the inspiring things that other youth are doing around the country really helps.

—Larry Kraft

At 6, my son does show some concern about climate change because he knows that I work on climate change and therefore we talk about it more than most families. He does know enough to understand that climate change is a problem and that we must transition from using fossil fuels to renewable energy. But he doesn't yet know enough to be really worried, angry, or afraid. Frankly, I am dreading the day when he turns to me and says that he is angry with me—and other adults in his life—who knew but did not act fast enough or bold enough to make sure he will have a safe and secure future.

—Jill Kubit

Regardless of their ages, do remind your children that, as a result of the crisis we face, there is an infinite number of opportunities that can and will help our planet begin to heal. As we discussed in Chapter 3, envisioning the future is a critical step to creating it. Once you can see it, talk to you kids about envisioning this future together. Regularly discuss the many ways and opportunities open to your child that will jump-start the creation of the future they want; this will help take away some of the fear and build up active hope.

My daughter is only 3, so she is not yet aware of climate change or what it might mean for her future. As her mother, however, I am acutely aware that her world is less stable, and the future less secure, than mine. I worry about her health and physical safety, and I worry about the society she will mature into. But, as her mother and as a mother, I also believe strongly that I have a responsibility to play my part in the global fight to course correct. I try to do this in my personal life by adopting climate-responsible behaviors and in my professional life by being part of an organization that works to increase climate and energy awareness in the media and across society. As my daughter grows up, I am also committed to helping her appreciate the world around her, to working with her to take climate-positive actions as often as she can, and—most importantly—to encouraging her to never be silent in the fight to protect our species and our planet.

—Rachel Arkus

In addition to, or in the absence of formal lessons and examples, our children are learning about our crisis from their friends and by watching and modeling what your family, your community, and your neighbors are doing; they are also learning about the crisis directly on social media. Because of this, it's important that they learn the facts about climate change, its impacts, causes, and possible solutions directly from you or from an educator you trust to share this reality. Too many of our children are being personally impacted by our climate crisis already. For many, there is no doubt about the reality of the climate crisis—they are living it. But by learning directly from you, at an early age—or in a formal educational setting, at any age—and by seeing other children and adults around them working to create a livable future where they will not only survive but thrive, your child can build hope and resolve. They will also begin to envision a way they, too, can work directly on the solutions puzzle.

We work together as a family to do our part. I remind them what goes in the compost, what goes in recycling. They can ride their bike instead of asking me to drive them, and we talk about voting and looking at candidates who take climate change—and my kids' future—seriously. They are diligent recyclers and composters. But they know it's not enough.

—Chrysula Winegar

Find Your Superpower

For Mother's Day, my kids hung a Wonder Woman poster up on my office wall. It reminds me daily that I have two personas. The first is Harriet—mom, wife, sister, daughter, neighbor, and friend. This is the person many people know and see around town—at the post office, the supermarket, attending a neighborhood party, or out for dinner with friends. But this isn't the full or complete me. There is the other Harriet, the one who speaks out at marches and leads climate chants from the steps of state houses and city halls, the one who serves as the police liaison at rallies or marches, and the person who assists on court support (being at court, or facilitating support for people who have been arrested for taking non-violent direct action). This is the also the Harriet who calls and visits her elected officials regularly, who pitches in at art builds that are used at rallies in city parks and at community centers in lower Manhattan, across New Jersey, and around the country. This Harriet regularly gives talks and moderates discussions on the realities of our climate crisis for anyone who asks. This Harriet also knows intuitively when two people should meet; she can sense when there will be a positive and important connection to be made that will advance climate action, discussions or policies—she has this superpower and uses it to amplify and create bigger actions that lead to climate solutions.

I need to remind myself repeatedly that I am one and the same person. Too often, these two personas seem separate and far apart. Many of the Climate Mamas and Papas I have had the honor to meet and spend time with often feel this same way too. At home with our families, we attend Thanksgiving dinner with climate denier Uncle Bob and argue with him—or ignore him as many in the family suggest: ...*this is family time, don't cause a scene.* At times, we hide the full force of our Wonder Woman persona as our own parents, siblings, and good friends regularly remind us not to ruffle feathers.

My advice, though, particularly with your own family, is to not shy away from or be afraid of difficult discussions at family dinners, special events, and holiday get-togethers. Your children are always watching what you do and what you say. Do listen to Uncle Bob, who may have a very different perspective on almost every issue than you do. Try starting conversations with him from a found place of common ground and shared values. Show your children you are trying. Show them you do not just dismiss or ignore people who disagree with your views or beliefs. To be successful in slowing down our climate crisis, we need people with a multitude of ideas, views, and political persuasions. Finding common ground on climate solutions is critical for our present and our future. While you may have family and dear friends that you disagree with politically, working together to remove climate action from the realm of partisan politics by putting the climate crisis front and center is something we all must strive to do.

Model for your children a conversation with Uncle Bob that begins from a place of love. Perhaps it begins with a shared and treasured family story. Make it personal, make your concerns and your hopes heard. At the same time, we mustn't normalize situations and actions that are clearly not normal, nor allow lies to be spoken and remain unchallenged. Everyone is entitled to their own beliefs but not their own set of facts. We must speak the truth. Work to uncover it, and then champion it.

My family is very engaged in my career as a legislator. In many ways my boys have informed my position on environmental policy and climate change. They attend community meetings with me; we discuss policy at the dinner table; and I am brutally honest about the challenges of getting climate change legislation passed in the state house. My boys get front row seats to my advocacy on this issue.

—Donna Bullock

Wonder Woman is always there—even if, at times, she operates below the radar. She often doesn't have to say anything. Her presence alone is enough. When she walks through her kid's school parking lot people turn off their cars and stop idling. At dinner parties, friends hasten to explain why the recycling is mixed with the garbage; they apologize and let her know it was a mistake. When a friend buys something that is organic, sustainable, or from recycled goods, Wonder Woman is the first to hear about it.

I don't think separating these two selves necessarily is a conscious decision. It is just reality and how life presents itself—in many settings and in many communities. Raising a family and the mundane tasks that this brings often may be at odds with the activist mom or dad persona that is constantly present but hiding in plain sight. We all find ways to combine the two at various times, often without thinking. Our children see through to who we are; they are always watching, even when it seems they are not.

Children intuitively know right from wrong. Their built-in moral compass detects and rejects injustice. And at a basic level, kids understand that the "good guys" can stop the "bad guys." My kids see me—their mom—as one of the "good guys," keeping them safe and stopping bad behavior. In a way it's an obvious extension of the things they see me do as their parent. I think it's comforting to them. My kids understand that part of being an adult is joining with your community and helping shape it. And like most kids, they aren't yet jaded and cynical. They believe it's possible to build the world of your imagination.

—Anna Fahey

Over the years, my kids have accompanied me to rallies and protests. When they were young and preteens, they seemed thrilled to go with me and to be part of the action—being with their mom in these settings was exciting and fun. I'm not sure they thought too deeply about the long-term impacts of what they were doing; being along for the ride, having a voice and a say when and where they could, seemed to be enough. As they entered their teens, they seemed less inclined to join. I was no longer quite as cool as I had been. Not many of their friends or their friend's parents were actively involved in the climate movement. I see this coming full circle today as they journey through college and beyond; they are rethinking who and what they are, who they want to become, and where they currently stand. As I listen to them express their views, their beliefs, and their vision for the future, I feel proud and hopeful. I know they were watching, listening, and learning.

Name What You Are Working to Save and Why

I walked in the footprints of elephants. *Really.*

I am so very privileged to have had this amazing opportunity when I visited Botswana in the summer of 2017. This was my first (and so far, only) visit to the African continent. It was a life-changing and life-affirming event. Being in the presence of so many amazing beings in their natural habitat—elephants, lions, zebras, wild dogs, giraffes, hippos, and antelopes, alive and well and in great numbers—was truly magical and the stuff of my childhood dreams. It also made me realize just how threatened these amazing creatures are and how truly precarious is the Shangri-La that Mother Nature has created. Habitat loss and extreme weather are already working together to wreak havoc on unique locations and ecosystems throughout the world.

Traveling across the world isn't a requirement for creating an equally powerful connection with species other than our own. Ivy, our family dog, was in many ways my third child—the one who would never leave home and who loved me unconditionally. At 14, she tired more often, yet she remained ecstatic to see me even when I'd only been out of her vision, perhaps just down the hall. When I felt particularly stressed, putting my forehead next to hers for a while, letting myself rest in that moment, connecting deeply with another species, relaxed me more than anything else could. I miss her dearly but will always hold the memory of this loving connection with another species deep in my heart.

Find time to be together with your children in nature. There is nothing like touching, feeling, and being in nature to remind us how precious and special our planet is and why we must strive to keep her healthy and safe. Even as infants, many of our children seem drawn to other species; they easily feel a connection, a love, and a fascination. As they grow older—in their own words and ways—they often express worry about species and biodiversity loss and the need to protect all things wild. Discussing nature, biodiversity and wild animals and interweaving stories with reasons and ways to protect our natural habitat and species other than our own is a powerful way to share the climate story in a non-threatening and motivational way.

It feels like this conversation has always been part of our lives since they were very little—I would say it began around age 3. Between home and school, my children have always been aware of the planet's fragility and our impact on it. It began with concerns about animals and plants facing damage or extinction. The massive loss of biodiversity has been something that all of my kids have cared deeply about, but especially my eldest son, who is now a newly minted teenager. He has—since he was very little—been obsessed with animal life and has consistently prayed almost every night of his life for those creatures facing extinction.

—Chrysula Winegar

Climate Curriculum

Many of our children, whether in preschool, elementary school, middle school, or high school, will encounter some type of formal lesson in their classrooms on climate change. Some of these lessons will be based on strong science. Many of these lessons will be well intentioned. Some of them will be well thought out. But too many will not provide scientific facts about climate change or effectively connect the broader dots to what is causing our planet to change. An important step you can take is to find out more about the climate curriculum being taught in your children's school district. What lesson plans are being used, what are these based on? Are teachers trained on climate science? Is climate change integrated into other subjects and lesson plans?

At a time when many teachers in the US and around the world are already not adequately compensated for their work, they are not regularly trained (and certainly not often re-trained) on the latest science of climate change. Nor are they updated or trained on the intersectionality and cross-pollination of lessons that could showcase how climate change impacts every facet of our lives and every subject of learning, including civics, economics, politics, health, social studies, and the natural sciences. If teachers are forced to search out their own materials and curriculum, they may choose not to do so, or they may not have the time to do so. Yet, many resources exist that can provide peer-reviewed curriculum that fits with nationally approved standards across subject areas on climate change. Is your school district making this easy for teachers? If not, can you help? Let your child know about the situation, and let them see that you are helping out on this one, too.

As an example, Young Voices for the Planet[1] is an organization that showcases and provides lesson plans on teaching *self-efficacy* as a way for children to model behavior that is hopeful, self-affirming, and effective. The organization has short films and accompanying curriculum readily available for teachers, from kindergarten through twelfth grade. Self-efficacy is the ability to get things done, to show there are actionable ways forward, even for a child. By providing examples for our children of other children who are already leading on climate solutions, we can help our children find their own peer role models, create solutions of their own, and have their voices heard.

Parent Climate Activists

Since the early 2000s, millions of people have marched and rallied in their state and national capitals demanding climate action. More and more people are concerned, active, and activated. But these numbers collectively, are still small. Marches, rallies, and protests are often noted one day and forgotten the next. As of the writing of this book (2019), there has yet to be a broad mass global or national uprising—in any country—around our climate crisis, much less one specifically from the parent voice. As I write this book, I feel we are on the verge of this happening. In September 2019, more than four million people marched for climate action for the world. By the time you read this, perhaps tens of millions will have been activated and this uprising will already have come to fruition. If not, I feel sure it will come in the near future.

Parents all over the world are organizing, in small groups and large ones, planning more directed, vocal, and visible actions; they are speaking out collectively and individually as protectors of their children's present and future. Solutions to our unfolding climate chaos are at the center of their demands. I see this unified energy on display in community centers, in church basements, in living rooms, and in the streets, around the US and the world. I am regularly on calls with parent organizations and grassroots (local level) and grasstop (professional or more established) groups that are discussing cooperative and collaborative actions and devising ways to mobilize and act. And in many cities, small and large, parent action is manifesting in a much more visible way through non-violent direct action in various forms.

According to the Merriam-Webster dictionary, *direct action is action that seeks to achieve an end directly and by the most immediately effective means—such as a boycott or strike.*

We are witnessing the growth of movements across many sectors and increasingly between sectors as well: the #MeToo movement, the tragic Parkland shooting and resultant youth mobilization on responsible gun safety, protecting transgender rights in the military, immigrant rights, standing up to hate groups, and simply but powerfully showing up in increasing numbers at the ballot boxes. Reminding ourselves and our children of the interconnectedness of these seemingly separate struggles is critically important. Since the Second World War, the Western, Developed or Northern countries—whatever label you choose to use—have experienced a relatively calm and prosperous period. This time is coming to an end, and the cracks are visible in many, many ways. Showing up whenever and wherever we can is one of the ways we will create just, equitable, and long-lasting solutions and a stable climate—the critical linchpin for all of these interconnected rights.

In the summer of 2018, Greta Thunberg, then a largely unknown Swedish 15 year-old, began *striking* each Friday in front of the Swedish Parliament buildings; she would sit there, on strike instead of at her desk in school. Her *Fridays for Future* individual protest has ignited an international movement of children and adults across the world, and, as I write this, it seems unstoppable.

Parents for Future; Extinction Rebellion—Parents; School Strikes—Parents; Sunrise Movement—Parents; and other, similar, grassroots and direct-action parent campaigns have sprung up in in the past few years. These groups have been spurred on by the dire scientific reports of late 2018 and by the actions, voices, rallies, school strikes, and urgent calls to action from our children. Parents of all ages, grandparents, uncles, aunts, teachers, godparents, and friends are all concerned and activated. Many of these new parent-focused groups are extensions of other activist grassroots groups that have become more visible and active since the 2015 Paris Climate Agreement and the more recent Greta Thunberg's Fridays for Future movement. These groups are, for the most part, focused on non-violent direct action.

US-based organizations like ClimateMama, Mothers Out Front, the Environmental Defense Fund's Moms Clean Air Force, and the Sierra Club's Climate Parents (in addition to a wide range of regional and local parent-led climate groups) have been working for years, laying the groundwork and the foundation for these new groups. Through parent education, best-practice examples, raising up of parent voices, and direct and non-direct action and activism, they have been building coalitions and providing parent-focused information on and about our climate crisis for years.

As collective action *online* evolved and gained success in the 2000s, much of the early, directed activity from parents around climate tended to be online. Nap-time activism was encouraged: click on a link while your kids are sleeping; share resources; learn what you can in the moments you have. These actions will and do continue, and certainly they have a place in the broad movement to educate and activate. However, these actions take second place to being directly present and visible.

Striking the flint over and over, early parent climate groups have been waiting patiently—and sometimes impatiently—for the spark that would light the fire. It seems the fire has finally taken hold. How big the fire will grow remains to be seen, but this planned burn has created a new forest floor, ushering in new life, which is growing quickly. In basements, school yards, backyards, and around the water cooler at the office, parents who have been worried about the climate crisis and unsure about what to do are sharing their concerns, talking more directly about our climate crisis, and finding more and more ways to come together to demand action and build solutions.

Setting up a recycling program in our children's schools (or at our offices) is a beginning, but it must not be the ending point. We must dig deeper than we have in the past and find new ways and spaces to consciously and intentionally add climate action to our list of parental responsibilities. Each of us will create a way forward that fits with the unique demands and flows of our own families. However, our intentions and actions must stretch us and push us further than we think we can go. While we live our lives and care for our families, there must be a conscious understanding that we are in emergency mode.

Our Kids' Climate[2] is one manifestation of the new parental call to action. It grew out of a meeting of established parent climate groups that united in 2015 to send a message to the United Nations, demanding climate action on behalf of our children. This parent coalition relaunched in the spring of 2019, with 18 parent organizations from around the world, and the group is growing and expanding weekly. Its goal is mentoring, educating, and helping parent groups around the world become established (without having to recreate the wheel) and sharing best practices that can be replicated and amplified.

My boys are 9 and 7, so they are quite young to be thinking about the crisis. Nevertheless, they hear about it [climate crisis] because I am working on it and they know I am so concerned about it I wrote a book[3] about it and I go on climate strike every Friday! They are lucky in a way because they are surrounded by many people who are taking action, and they join me on protests when they can. When they say they are concerned, I tell them that there are many grown-ups who are doing everything they can to solve the crisis. They see me and others trying to show leadership on this, so I guess they believe me when I say that. I also tell them that when I was growing up we had a major crisis that worried us too. In my case, it was the fear of nuclear war. We really thought that the big bomb could come at any moment and that we would face a nuclear winter. I say this to acknowledge their concerns but also to allay their fears—the threat was real, and we were scared, but people worked together to build peace and it worked. We are still here and the nuclear threat is less, if not gone. We have to work together to solve the crisis rather than focusing on all the bad things that might happen. I think it is really important to help them see that there is hope we can turn this around.

—Lorna Gold

There are no magic words, secret handshakes, or singular actions that will guarantee that the world will be a safe place for our children, or that we can create a future that will be better than today. Anyone who promises this isn't telling the truth. However, the words of wisdom and advice freely shared in this book, including those of dedicated, committed, and trained climate educators, scientists, communicators, and activists, should provide you with some solace, as well as clear and replicable paths and ideas that can take some of the fear and the uncertainty out of your next steps. These stories and examples can help equip you and your kids with some of the tools and skills that will help you manage many of the near-term complications that our climate emergency is already bringing to us. In the next two chapters, we will look specifically, by age group, at ways to introduce climate change into family conversations and at ways to engage children at different ages and at different levels of activism, on climate action.

5

Transitioning from Angst to Action

Preschool through Middle School

Children under 5, in almost all instances, are too young to even begin to comprehend the precarious world they have been born into. Upon bringing a child into the world, your immediate task is to keep them safe, well cared for, well fed, and well loved. Starting when they are babies, your love and concern for your children will be shared in large part almost as if through osmosis; as you talk to them, read to them, and are truly present with them, they will feel your love and know deeply that you care about them. As we live climate change—as life itself becomes more challenging in so many ways—feeling loved and giving love takes on added importance. If you are a hugger, create more opportunities to give your child hugs. If you are not, show your love through your looks, words, actions, and deeds and through time spent together.

CLIMATE MAMA

Our oldest daughter, who is almost 5 years old, is too young for us to tell her that the climate crisis is a crisis. However, that does not mean that we do not talk about climate change at the family dinner table. The way we introduce the topic is through trying to teach her both a sense of wonder for the natural world and a sense of responsibility to care for it. For instance, when using resources such as water to brush her teeth or paper to do her art projects we will tell her that caring for the river, the trees, or the birds starts by not taking more than what we need. If she turned off the tap when brushing her teeth, we would say things like "That makes the river so happy that you thought about turning off the tap."

—HELENE and RAOUL COSTA DE BEAUREGARD

The Climate Conversation: Preschool through Middle School

The 2018 United Nations Intergovernmental Panel on Climate Change Special Report, *Global Warming of 1.5°C*, definitely lit a fire for many of us, young and old, at the same time that it scared and silenced others. For some of us, it is the stuff of nightmares. So, if your children are up at night worried about our climate crisis and the end of our world as we know it, do remind them that many, many people are concerned too and are also working night and day on solutions. As a parent, sharing the truth about our climate crisis is key, as is sharing the reality that we don't know for sure how quickly actions can begin to help our planet heal, once put in place. The urgency remains and is something we will live with, but it is carried on the back and in the arms of us all. The depth of your explanations on the reality of the crisis, as with many things you discuss with your child, should be geared to the age of your child and to what you know your child can handle.

When my daughter was four and a half, she asked me point blank about climate change. You'd think I'd be equipped for this conversation. For a decade I have studied climate messaging and run workshops on how to talk about the science and policy solutions.

Looking into her eyes in that moment, I wept. I was speechless. It was the first time I'd cried about the climate. My daughter had cracked open a heavily guarded vault of emotions.

We started out exploring nature and talking about science, weather, and humans in the environment. We observe the world around us and talk about what we see, from birds and bugs to buses and bikes. Kids know stinky exhaust from fresh air. Pollution is a toddler-level concept! And kids are naturally curious, observant, and creative—they can get excited about nature, science, and new ideas. It's more about awe than fear when they're young; my 5-year-old asked me the other day, "What's your favorite natural disaster?" The politics and the seriousness of climate impacts come later, when kids are a bit older. But if you think about it, dealing with climate change is about things kids already know well—the values we work hard to instill from the time we become their parents and the ideas they learn in preschool! It's about cleaning up our messes; about the power of the sun, wind, air, water, and our own bodies; it's about treating all people with respect and dignity, about stopping bullies; about sharing; and about making rules that keep us safe—and making sure everyone follows the same rules!

—Anna Fahey

For very young children—children in preschool and lower elementary school grades—make time to read, share stories, and be together in nature. Hearing about children and adults working together on climate solutions and people taking action, doing good work to protect Mother Nature and all her creations, is important for all of us, regardless of age. For our youngest children, these stories can be shared through verse, song, dance, books, and simple words and by attending climate-focused events together. Even if it sometimes feels as if your child is too young to fully grasp what you are sharing, moments and memories around climate action can be spoken about years down the road.

I honestly can't remember a time when we did not talk about climate change, in one way or another. I've always tried to make this age-appropriate, and so we started by talking about the importance of protecting the Earth and making sure we had clean air and clean water. When he was about three years old, we took him to his first climate change action to protest against oil trains. We carried a small stop sign from red-and-white construction paper and chanted "stop dirty water."

It became so commonplace to talk about how mama was working on climate change that when he was about 4 years old, he was completely surprised to find out that I was against climate change. I recall that he one day said something along the lines of "but you like climate change, don't you?" It was then that I realized because we had only talked about how I worked on climate change (but hadn't yet defined it), he thought it must be a good thing. It was at that point that we began to tell him a little bit more about climate change, but we were definitely cautious about our language. It was also during this time that we began to talk about solutions. We'd point out solar panels and windmills, talk about why we started composting, and discuss how lucky we are to live in a place where we can walk, ride our bike, and take the bus or train to get to the places we needed to go.

—Jill Kubit

Sometimes, words and stories create themselves. Sadly, as we live climate change, extreme weather and climate impacts will become more common occurrences in all our children's lives, through their personal experiences or experiences of family or friends. These events will provide opportunities to discuss climate change directly and should be addressed carefully, in an age-appropriate way, with your children at every stage of their lives.

When our son had just turned 7 and our twin daughters, Anya and Isabella, were 5 years old, Superstorm Sandy hit our New York City hometown—and our children's understanding of the climate crisis—hard. When our children saw homes and structures washing out to sea and spent nearly a week with large parts of New York City flooded, without power, and with schools shut down, all three began to understand that climate change wasn't just something that was discussed a lot or debated but that it was clearly something that they were living with and experiencing for themselves. Fortunately, while living with and witnessing the continuing impacts from Sandy has been challenging for our children, it has not been overwhelming or depressing for them. Being able to open our home to friends and colleagues for hot meals, sleepovers, and showers during the Sandy Blackout helped our children begin to see how a community can work together to take on climate challenges.

—Allan Margolin

Creating a family climate plan or family climate roadmap is a great family exercise that can be expanded, adjusted, and discussed throughout the years, on a regular basis. Make your plan simple but straightforward; it should be a positive way to keep the climate conversation a part of your family's everyday lives in a proactive way. Perhaps you already have regular family meetings as part of your family time together. Why not add a climate component to these regular discussions?

Over the years, we had conversations about which products have palm oil, our personal consumption of fossil fuels, and more. Most of these conversations they initiated and took the lead on. We researched questions together. We discussed my work as a legislator, and specifically my community solar legislation. They shared lessons they learned in school or on field trips. We watched every Disney and Nature movie released on Earth Day. We participated in neighborhood cleanups. We learned about the climate crisis together.

—Donna Bullock

If you don't have regular family meetings, there is no time like the present to initiate them. Organizing around climate actions can become a part of your family routine and your everyday conversations. What is your family doing, in the short, medium, and long-term to address the climate crisis? This can be discussed both from the mitigation side and from the resiliency and the adaptation side.

Our young children are still too young to fully comprehend the climate crisis, but my wife and I do our best to be willing to speak the truth about consumption and waste, about how some cars burn gasoline that comes from oil and that these pollute vs electric cars. We spend time together in nature, pointing out its beauty, strength, and fragility as well. More recently, [my 5-year-old son] Trevor has occasionally asked us if it is the end of the world. If we lived in an ideal world, I wouldn't have to think twice about how I respond.

—Derek Hoshiko

Mitigation

Explain mitigation to your children. For them, it can be translated as lowering your greenhouse gas emissions as a family. Some ideas include meatless Mondays, lights-out Tuesdays, composting, rain gardens, and tree plantings. Your family can think together of ways to mitigate the crisis by lowering your family's greenhouse gas emissions. Talk to your kids directly about coming up with solutions together.

My son is not the type of person to ask a lot of questions initially, but he internalizes things and thinks pretty deeply about them. He is a vegetarian. Initially, it was because he didn't like the taste of meat. He also loves animals and nature. When he learned where meat came from, he decided that he didn't want to eat animals. We went to an exhibit on climate change in San Jose, Costa Rica. Part of the exhibit discussed the impact of raising cattle and eating meat on climate change. He was really interested in the exhibit and spent a lot of time reading all of the information. We could see he was really thinking about it. Well, the next day, he said to us, "Mama & Daddy, I'd like you to give up meat for a week for the Earth." We were floored, but you don't say no to something like that.... So we negotiated a bit with him, and our entire family gave up meat for a month.... While we have not all become vegetarians, our eating habits have dramatically changed since then because of Jason.

—Larry Kraft

What's your family's carbon footprint? If you don't know, you can't reduce it. Research and find an ecological calculator that works for your family and fill it out together so you can understand your own family's footprint and reduce it.

When we drew our formal Family Climate Roadmap, we wanted our daughter (4 years old) to participate. She was really excited about the project. When building the roadmap with her, we did not discuss why we needed to reduce our emissions as fast as possible beyond saying that the trees, the birds, the animals do not like summers that are very hot. (We live in a place where summers are [or rather "were," until recently] usually mild.) Instead, we focused on what we would do to help them. For instance, we discussed how we would make our meals extra yummy by adding more vegetables, grains, nuts, and fruits (i.e. moving toward a plant-based diet). For each of the categories in the Climate Roadmap we discussed (transport, housing, diet, goods, services, past emissions), our daughter drew columns with a picture representing those categories. She was really proud and now is talking about sharing the plan with her friends!

—Helene and Raoul Costa de Beauregard

Resiliency and Adaptation

The resiliency and adaptation side is about how we can build up and protect ourselves from our unfolding climate crisis, and how we want to adapt our lives to deal with the changes that our climate emergency will inevitably bring. In all likelihood, you have a fire emergency plan and a safe place to go in a natural disaster, with emergency numbers and more. While there is nothing natural about what we are doing to our planet, talk about climate disasters with your children: how we are making hurricanes worse; how when it rains or snows, it does so in extremes; how it's hotter in the day time and cools off less at night then it did when you were young; how allergies are worsened, and on, and on. Sharing the science and the impacts of our climate crisis with our children is important for their understanding, even at a young age. Overtly planning for climate emergencies should also be part of your emergency plans as well. As families, we must acknowledge and treat our climate crisis as the emergency it is. We must prepare and be prepared.

Regardless of whether you live in an urban, suburban, or rural location, I am sure you visit local parks, the public library, zoos, the beach, campgrounds, or local community gardens with your young children. They are never too young to discuss conservation, restoration, and nature. In his book *Last Child in the Woods*, Richard Louv reminds us of this: "Children need nature for health development of their senses, and, therefore, for learning and creativity."[1] You may not know it at the time, but your children will learn and pick up knowledge just through exposure to the special places and people you love. If you don't already, start looking in your local paper, community parent websites, and town event calendars for family-friendly programming and activities.

I can't pinpoint the initial conversation. It is something we started to talk about early on—mainly because of their interests. My boys were probably 3 and 6 when I first overheard them discussing climate change. It was their first "argument." My younger son wanted to turn the light on in their room, and my older son told him they didn't need the light because it was daylight and the sun was shining in the room. The older son further explained that we need to conserve energy to save the polar bears. I assume they learned this lesson on one of their many trips to the Philadelphia Zoo.

—Donna Bullock

Engaging Your Child on Climate Action: Preschool Through Middle School

The climate crisis and democracy crisis are interwoven. Climate change inaction is a social science, psychology, democracy, and communications problem. Our goal [Young Voices for the Planet] is to empower youth and help them develop the self-efficacy necessary to become change-makers and a force to motivate and encourage parents, communities, and politicians to make the necessary changes to demand a functional democracy and protect our planet so that it can continue to support the survival of future generations.

—LYNNE CHERRY, Founder of Young Voices for the Planet[2]

While many of our children will follow our lead, others will lead *us*. Even when you might think they are too young to do so, let your child lead if she wants to; she can and probably will surprise you. But be there, all along the way, to help her navigate the twists and turns that will arise.

My most important role in life is being Anna's Mom. Like most mothers, I am driven to protect my child at all costs. This is why I find myself parenting a youth climate crisis leader. I find myself in the uncharted parenting territory of youth climate crisis activism. Fortunately, I can think my way through these sometimes murky waters, find the hope, and take action. I live with constant apprehension that my daughter isn't emotionally and psychologically equipped to deal with an endless parade of interpersonal and

> political ugly that comes with this political territory. However, I have determined that Anna's activism is in her best interests because she is the direct beneficiary of her efforts.
>
> —KATHY MOHR-ALMEIDA

As more and more children find role models in other child climate activists, we as parents must continue to build and share our own experiences and knowledge in supporting our own and other youth climate activists. During the September 2019 Climate Strikes, Our Kids' Climate[3] offered the following three ideas for parents as ways to support their children:[4]

1. **Social Media Support:** Use Facebook, Twitter, and Instagram to elevate kids' messages and stories. Post an image or story on social media about how you support climate activist kids, or repost something from a youth striker that you are inspired by.
2. **Lead Others in Your Workplace:** If you can, strike from your day job, or at least walk out for a short time. Find a strike near you. Talk to your colleagues about why you are doing this. Do remind your child (and yourself) that many parents around the world will be unable to attend a strike or walkout. Consider organizing a strike event near your work or in your community, invite a youth climate striker to do a talk, or host a lunch-time discussion about the climate crisis with your colleagues.

   Sign ideas:
   - Concerned Parent/Aunt/Grandparent of child(ren)'s name(s) and age(s)
   - Fight for Their Future
   - Parents Unite for Our Kids' Climate
   - Your Future is in Our Hands
3. **Talk to Your Child's School:** No matter what grade or level, let your child's teacher or school know that you care deeply about our climate crisis and that you support the youth strikers. In an email or through personal conversation, encourage teachers

and school leaders to support the strikers' actions, talk about the youth-led strikes in their classes, and integrate climate change into the curriculum. Or, why not strike up a conversation with other parents in the parking lot or at the school entrance about doing something together?

The increasing frequency of climate strikes and other intergenerational climate actions can help you elevate the discussion around our climate crisis with family, friends, neighbors, and strangers. We, as parents, will find ourselves continuing to refine, create, and share parenting tips amongst and between one another as we live life in the Anthropocene.

In this regard, if your child signals she wants to take on a leadership role in her school, your family, or at other neighborhood events, and as she learns and becomes more engaged by watching other youth who are leading, support her, but don't forget to remind her that she and other activist children are still children. Remind your children that there are many, many adult allies working with youth across generations. We have been building a foundation of climate activism for decades, and we will continue to work on this foundation and make it stronger and stronger—for decades to come. The present and future does not rest on the shoulders of your child alone. Make sure he understands this clearly.

I usually frame it as "we've got this"—the larger collective "we." I show him all the other kids around the world who are a part of Fridays for Future. I share the actions of organizations like Extinction Rebellion, Sunrise, This is Zero Hour, and 350.org.

—Eve Mosher

At the same time, your kids are never too young to accompany you to a march, a rally, or a climate-focused event. Oftentimes, these events will have a family-friendly area where young families will gather together. This will likely be listed as part of the march/rally activities. If this area isn't designated, perhaps you can organize one (if there is enough time and if you can reach out to the event organizers). Otherwise, look for other families with kids your kids' age attending the event. Perhaps your child is in a stroller, on your shoulders, or walking hand in hand with you. Find others in similar circumstances; walk, stand, rally or march together. You and your children can meet new friends and be part of the climate action networks that are building in your area. Or maybe you and your children will be the ones to start building this network.

I was really encouraged by pictures I saw from other parents—especially in Australia—going to rallies with their preschoolers. So, I went to the March 5, 2019, international youth strike with my eldest daughter (5 years old). I thought it was important for her to have such an experience and see that there are many people who are caring for our planet. I also wanted to show politicians that there are many parents who are as concerned as the students striking and who, unlike them, can vote and will vote for the future of their children. I knew the rally would be short and peaceful, but it was also important for me that the event be a joyful and fun experience my daughter. We drew together her size-adapted sign focusing on something she could relate to, superheroes—Be A Climate Super Hero: Save Our Planet—with nice pictures around it. I brought some snacks and packed a few books, pencils, and sheets of paper so she could draw when seated on the grass listening to the speeches. I march with Fridays for Future every Friday, but I do not bring my daughter. I want her to be older and make the decision for herself. I borrowed from the playbook of Madres por el Clima and Heidi Edmonds from Australian Parents For Climate Action, who made special t-shirts or capes for their young kids to wear at preschool or kindergarten. With the help of my daughter, we took a white t-shirt, a few color pens, and created her #ClimateTeam t-shirt. She wears it to school when she wishes to.

—Helene Costa de Beauregard

As the climate crisis worsens, more and more of us need to be seen and heard. Let your child guide you; they will know when and if they are ready to lead.

In my experience, when Anna leads the way, she isn't pushing limits she isn't ready for. If your child is 6 and is worried about the polar bears, talk about it and help your child devise an intervention, no matter how small and seemingly impotent, because you never know where this small step will lead your child. Anna started her activism, unbeknownst to me, in 2nd grade by making jewelry out of trash and selling it to the parents at her school. She made 80 dollars and immediately sent it to one of her pet causes.

I don't believe anyone can evade what they can't see. However, asking deeply reflective questions about my motivations and what is shaping the behaviors of others usually proves prophylactic. For example, I am aware that a part of Anna's effectiveness is her young age. Anna's activism won't always be a novelty, and the window of attention because of her age will diminish. My husband and I encourage Anna to actively explore interdisciplinary fields so that she has many doors to walk through, if and when her activism becomes less central in her life.

Perhaps the better question is how much more confident will Anna be at 18 compared to her peers who spent their adolescence in other ways. Whatever Anna choses to explore as

> her professional path, she will be better equipped in that field because she will have more diverse life experiences than many of her peers.
>
> —Kathy Mohr-Almeida

Some children will not be ready to actively lead, and your job as a parent is to protect them so they don't get pushed to do so before they are ready. Many others likely will not be capable or interested in being a climate activist leader, ever. But even if you know that leading isn't right for your child yet, it doesn't mean that you should stop showing him how you or others are leading. Make sure to point these moments out to your child so they can watch and learn from you. Also, when you see young people taking a lead role, make sure to point this out, too. Share your thoughts, hopes, and feelings about how people of all ages are taking climate action. Also remind him that each one of us can be a leader or a team supporter in different ways and at different times in our lives as well.

> To be honest, Zayne has known a little something about climate change since he was 4 years old, drawing future flood lines around Miami [as part of my art project]. But I don't think he remembers that too well. He got active in late 2018 after hearing about climate change (again, it wasn't the first time) on a kids' podcast, and he started feeling anxious. Now that he has gotten more active and learned more, we can really talk about the politics of climate change, what it means to take action, what makes good policy and how to combine personal action with supporting smart policy, and even understanding direct action. We talked a lot about the really smart, passionate people working on the problem. Since he has started taking action (after reading about Greta and her strike), he has gotten more savvy about the

reality of the situation and the need for leaders to step up to the challenge. I think he is both frustrated (sometimes it seems like nothing is happening) and buoyed by being part of a larger movement. In addition to the Fridays for Future weekly and big strikes, he has been speaking about climate change in more venues and participating in Extinction Rebellion actions. I think he is pretty aware of how hard this change is (we sometimes sit and talk about what the world could look like if we decided to make a change), but he is a fighter. We listen to Greta's speeches and those given by other youth at events and he is interested in understanding the policy proposals and trying to figure out if they are enough. He's mad, for sure, and honestly he has (they have) every right to be, but he is watching, listening, and learning how to channel that anger and frustration into action.

—Eve Mosher

Over family meals or during quiet times together, chat about the actions you have seen or participated in. Use these opportunities to discuss the importance of justice, fairness, equality, and equity and how interconnected and important these seemingly separate concepts are to creating climate solutions. Also, action doesn't have to mean striking, marching, or risking arrest. Climate action can be writing a letter, drawing a picture, putting on a play, talking to a neighbor or family member, planting a tree, organizing a neighborhood plastics collection drive, or designing a climate plan for school or home and sharing it with others.

I think because I work on the issue so much, the climate crisis has always been present in our family life, right since they were born. My boys shared the stage with me at their first climate protest in 2015 when they were still tiny. I never sat them down and chatted about it. We have a great story book about dragons and climate, *The Trouble with Dragons*,[5] which we still enjoy reading together—it helps explain how our lifestyles are destroying the planet. It is done in a child-friendly, fairy-tale–like way, and designed to encourage them to think. They are quick to point out when people are being "climate dragons!" I like the fact this book gives a child appropriate language that both acknowledges the root of the problem and the solutions. Once I sent it into school, as my older son told me they were asked to bring in books about the climate. He was very confused when the teacher sent him home saying "We are looking at the climate, not climate change!" As they get older, I will encourage them to find out themselves more about what climate change means, but for now I think it is really important to transmit ways to empower them and identify the behaviors of the "dragons."

—LORNA GOLD

Facing Fears

Finding ways to share how we are working to tackle our fears, our concerns, and worries head on, will help our children tackle theirs. This will take many forms and involves as many questions as there are answers.

CLIMATE MAMA

My self-preservation impulse is to pack away my own feelings, but as a parent I've become more conscious of modeling more open and honest ways of processing emotions. That's just one way kids can make us better adults. Instead of bottling up that anger and fear, we can tap into it and find ways for it to connect us to one another and also fuel our actions to make change. I want it to empower my kids, not shut them down. With my young kids—5 and 9 years old—I try to pull out, talk about, and relate to their feelings and share my own fears. The climate emergency is scary and big. What's particularly frightening is that it feels out of our control. I try to reassure my kids too (and reassure myself while I'm at it) by reminding them that there are armies of smart people, scientists and engineers and policy people like me, who are busy figuring out solutions. And like most adults, kids get excited about solutions—about innovations in energy technology and harnessing wind and sun to power our communities. My 5-year-old and I like to imagine the oil refinery near our town transformed into a wind farm.

—ANNA FAHEY

Remember, too, to ask your children what they think, and learn from them too. The wisest and most honest and direct answers often come directly from our children. They see things clearly; they are not yet jaded or influenced by nuances and prejudices.

Our kids were 6 and 8 when we first discussed climate change with them. One of the main reasons we did was that we took our kids out of school for a year and traveled around the world learning about climate change and other environmental issues. We partnered with a non-profit called the Wilderness Classroom, and we were sharing the things we learned with their middle school and elementary school followers. Those early conversations with both my kids were quite interesting. When we explained climate change to my daughter, she had a lot of questions. One was if we could do anything about it. I told her, yes, we can. And she said, "Well, Daddy, let's just stop it." Simple, direct, and for me, an emotional gut punch. Naïve? I don't think so. Her simple statement, which still drives me today, could be a key part of the answer. Passionate youth can connect with people and inspire the will necessary for aggressive action. Plus, we know how to address climate change. The solutions are available, we just need the will to do them.

—Larry Kraft

When my girls were young, I think their idea of climate change was really fuzzy—it was something they didn't necessarily understand as a concrete thing. They understood pollution but maybe not the connection to climate. Then as they've gotten older, especially my 11-year-old, I can see her making the connections between the industries and pollution we see in Newark and the threat of climate change. She asks more questions about where things are made or what we can do, especially about all the plastic in the environment. She understands more viscerally the injustice of climate change and the connection between the companies profiting from destroying the world and the communities that are dumped on all the time. When my girls tell me they're worried about what will happen to our planet because of climate change and pollution, I usually remind them of all the exciting actions that are happening all over the world and even in our community to fight for clean air and a healthy place to live, work, and play. I remind them of the marches that the children in Newark did in the cold and snow to protest the incinerator and the marches we did together in Washington, DC, to demand climate justice. I point to the thousands of people all over the world who are raising their voices to make sure that together we can fight back against polluters and demand action on climate change. This means there's still hope, and if we join together we can make a difference and protect all the places and people we love from climate change.

—Ana Baptista

If you aren't sure about how you feel about the many interconnected parts of our new whole, take time to teach yourself too. Climate justice issues are central if we are to be successful in finding lasting solutions and ways to slow down our climate emergency. We must take everyone with us along the way, and our children, from a young age, are sensitive and open to understanding this reality.

In our family, we've infused an environmental and health literacy theme into discussions of race consciousness, economic inequities, and other topics of intersectionality. Rather than skirt tough topics, we've tried to take a developmentally appropriate tack—like in sailing, where you move upwind by zigzagging. We use our surroundings as starting points for discussions. For us in New York City, that means doing things like expressing our gratitude for the vast public transportation system, noting differences in the transportation infrastructure and other built environment aspects when we travel outside the city. It also means naming the structural racism (for me this has meant a new education for myself about the historical context so missing from my own learning) that underlies the health and economic disparities and segregation so visible across our cities' neighborhoods and schools. As for when, these conversations have been happening for as long as we've been having conversations. For example, even before pre-K, we made a game of naming the origins of things: plastic, for instance, led to a discussion of fossil fuels. Fossil fuels led to a discussion about our electricity and heat. In this way, we teach the wonder and interconnectedness while simultaneously conveying a sense of stewardship, responsibility, and understanding that almost everything we observe results from human choices. It is the responsibility of those with power and privilege to have a perspective that these choices matter and have far-reaching consequences.

—Perry Sheffield

The Middle School Years

At the ages of 11, 12, and 13, most of our children can handle more detailed science information and more complex political and societal information. In middle school they should be digging deeper and more directly into chemistry, physics, and biology and therefore can more easily understand and connect the dots to the role humans are playing in causing our climate to change. In middle school our children are learning about society and the inequities that exist around the world, including in their own communities. As such, our kids should be better equipped to comprehend the climate injustices people face around the world. Even if these discussions aren't happening at school, they can happen in your home. So, this is an important time to educate yourself on what, if anything, your child is learning about climate change at school, and where and how the climate crisis is being taught at their school. Is the subject strictly taught as a science lesson? Is it covered in social studies, in language, in the arts? Start with your kid's teachers and the school administration, and if you aren't satisfied with the answers you are receiving, contact the local school board. Two good places to start getting educated and collecting information include Schools for Climate Action, which provides tool kits and background information for students and parents, and the National Center for Science Education, which defends and supports the teaching of climate change in schools.[6]

The middle school years can often be the most challenging school years for our children—on many levels and for so many reasons. Their bodies, their minds, and their ideas are evolving and changing rapidly at the same time as they are regularly being challenged by their peers and by their own development. At this age, friends can begin to take on a more prominent role in your child's life; your direct influence on what they do in their spare time may become more limited.

In middle school, your children will likely still need you to take them to their events, their clubs, their sporting activities, etc. If you were involved with these activities earlier on, it's likely you will stay involved in some fashion. If you child is interested, explore together ways to bring sustainable practices into after-school activities: are their school sports teams using reusable beverage containers? Does your child's school have an Earth Day celebration? Does the school regularly do collection drives for recyclables—traditional or non-traditional? Is there an environment club? Help your child come up with creative ways to advance sustainability practices that build on their personal passions.

My family's first exposure to the climate crisis was in 2006, when the entire family went to the movies to watch Vice-President Al Gore's first documentary, *An Inconvenient Truth.* We live in South Florida, and it became evident that our region of the state was at the highest level of risk because of the climate crisis.

At the time, my children were 8 and 12 years old. It affected my son, the 8-year-old, the hardest. It was not my intention to frighten him, but that is exactly what happened. So, our family decided to make some immediate changes to alleviate his fears and let him know that he could make lifestyle changes to help the situation.

Some of the steps we decided he could take were these:

- He and I started riding bikes to and from school.
- He started a Stop Global Warming Club at his elementary school. He taught the club members about CFL lightbulbs (before LEDs were preferred), planted trees around campus, and started a recycling program.
- In 2013, we both attended a Climate Reality Leadership Corp training to learn more about the climate crisis and learn how we could teach our community about solutions. He gave Climate Reality presentations to kids around town at our local library and at an Earth Day event.

—Monica Mayotte

Keeping Children Involved at All Ages

There is an endless number of things we each can do and countless ways to plug in. And we will need to keep plugging in for the rest of our lives. The reality that we will be living climate change for the rest of our lives is difficult to fathom and often too exhausting to think about even for us as parents, let alone for a young child. Our planet is a huge and complex system; she can't right herself overnight, yet we must continue to try to help her find her way, at the same time as we find our own.

So, as you help your child understand what we face, do grab one of the many books out there on *what you can do to save the planet*. These will offer you some great ideas that are tried and true to share with the children in your life as you build up your own family plans and roadmaps. From new recycling systems for your home to shared neighborhood community solar systems and rain gardens, activities and projects discussed, created, and worked directly on with your children and others in your community will help them understand implicitly how they can be part of the solutions as young children, as teens, and as adults.

Handbooks and advice on being a parent to your activist child, however, are fewer and farther between. This is one of the reasons I wrote this book. You can help us update our ClimateMama database on how to be a supportive parent by sending your ideas to ClimateMama.com.[7] Broadly, my best advice is to follow your parenting instincts. Keep your children safe even as you let them stretch their wings, find their voices, and express their feelings and emotions. Children have a unique way of speaking truth to power. They call it as they see it when it comes to injustice, inequity, and hope; and people listen.

The old adage, *children should be seen, not heard*, no longer applies. Whatever stage your child is at in their life-long learning about climate, when they feel doubtful that our personal steps are not enough to create the systemic changes required, remind them of Greta Thunberg and other youth leaders, and of the millions of children who are taking action and who have been inspired. When the time is right, building on a strong foundation, one person can launch a million voices and actions. As we live climate change, today is always the right day to create and support action on our climate crisis. Start now!

6

Pursuing Passions through the Lens of Climate Action

High School, College, and Beyond

ClimateMama blog, May 2016

I want you to know how sorry I am for the burdens my generation has placed upon you. We are forcing you to rise up before you may be ready, demanding that you become leaders and truth-tellers even while you are still figuring out who you are. We have taken time away from you. The existential threats we face and which we have created block a clear and uncluttered way forward, yet you are urgently tasked with finding a way.

I shared these sentiments when my children were in high school. Although scientists are plainly telling us we still have time to protect our children from the harshest realities of our unfolding climate crisis, time is running out. Being brutally honest with our adult children about what we face and that they have no choice but to face it too is part of our continuing role as parents. Recognize, however, that your children may not have the knowledge you do; nor may they be as focused as you are on the truth at hand. As they wake up to the realities of our crisis, guide them in both subtle and direct ways so that they can channel their own angst by finding ways to pursue their passions through the lens of climate action. Just as you are engaging more on our climate crisis—keeping your eyes open even as you want to look away—your children, as young adults or as adults with children of their own—may also just be finding their way. As their parent, you can help guide them.

As a parent, this is your job.

How to Share the Reality of our Climate Crisis with Our Grown Children

By sharing your journey on climate education and activism with your adult children, you will inspire them. As always, make sure to provide the facts and speak the truth, but encourage them to do their own research as well. Then, as they ready themselves to talk more about our climate crisis with their own friends, they will be more aware and up to date about the science of climate change and, hopefully, more savvy about how to communicate in a non-partisan, open and direct way. In this regard, suggest they look closely at studies by the Program on Climate Change Communication from Yale University. Since 2005, the Yale Program on Climate Change Communication has been conducting scientific research on "public climate change knowledge, attitudes, policy preferences, and behavior at the global, national, and local scales." As part of this research, the American public has been divided into six unique audiences—the Global Warming six Americas: *alarmed*, *concerned*, *cautious*, *disengaged*, *doubtful*, and *dismissive*.[1] Likely, many of you will fall into the *alarmed* category, along with me! However, don't be surprised if you find that your adult children are in a different category. Once their position becomes clearer to you, you can let your approach be tempered by their current perceptions and knowledge base.

The biggest question I've faced from my kids is, "Is it really that serious?" No kidding. They've grown up in such a different world from us, where everything is extreme (terror, the weather, gun violence, racism), at some level they think what's happening is normal. Also, I saw their attitudes change a lot after Donald Trump was elected, especially since Hillary Clinton won the popular vote. They really think the system is rigged and that people are generally corrupt (which is so sad) and that their individual action doesn't make a difference.

—Diane McEachern

As parents, we regularly encourage our kids (of all ages) to pursue their passions, to find meaningful work, and to try to make the world a better place than they found it. There are myriad ways to do this, just as there are thousands of career paths and jobs they can pursue. Whichever path they choose, you can encourage them to be thoughtful about making the world a better place by working on and incorporating solutions to the climate crisis in their plans.

My own kids—and the young people I work with in my job—all know my story. They know I gave up my career in journalism to take a leap for planet Earth. I am showing them every day this one clear action: devise a career path for yourself that is about the restoration, not the destruction, and do your best work.

—Jim Poyser

Here are several suggestions and conversation-starters to try with your adult child.

Ask them their thoughts about divestment campaigns. Have they thought about energy infrastructure and how it is built to last a lifetime? Remind them as well about the importance of taking time to say thank you, I love you, I appreciate you. As a follow-up to these conversation-starters, and to dive in with a little more detail, you could engage them on these topics:

1. Start a conversation about savings and financial planning. Do they have a pension plan through work or investments they have made themselves in mutual funds or stocks? Encourage them to talk to their broker or pension fund manager about divesting those funds from fossil fuel companies and their supporting infrastructure. While many of us may not have our own portfolio of investments, many of us do work for unions or companies that have established pension funds that are being invested for us. Jobs with the federal government, state and city offices, school districts with teacher unions, and many trade unions often include a pension fund as part of the contract. However, the manager of these funds, like many of your own friends and neighbors, may not be educated about the climate crisis, or she may not be looking at investments through an environmental, social, and governance (ESG) lens. More and more, these types of filters are being placed on our publicly managed funds. As these filters grow in popularity, companies will need to adjust to meet participants' expectations and demands. (As you encourage your adult children to do this, make sure you follow your own advice!)
2. Encourage your adult children to call on their state government to enact a *moratorium* on all new fossil fuel infrastructure. This includes pipelines; gas, coal, or oil power plants; and compressor stations and, of course, any new drilling or mining of fossil fuels. As we make the transition to renewable energy, existing infrastructure can and should be hardened, but building new fossil fuel infrastructure is simply a bridge to nowhere and a path down the road to climate chaos.
3. When someone they know takes action to protect our environment and our climate, remind them to thank this person for

helping create a more hopeful future. Climate change isn't just an issue; nor should climate action be identified as a mindset. Our climate emergency is a pervasive reality impacting every facet of our lives. Also remind them that while the climate crisis is certainly connected to the environment, it is much more than an environmental issue; your kids can help educate others by pointing this out at every opportunity.

Supporting Climate Activism

A September 2019 poll by the *Washington Post*,[2] conducted with American teenagers, found that the majority of our teens understand that humans are causing our climate to change and that they see themselves as directly in harm's way. The poll also showed that 1 in 4 students surveyed have taken some type of action on our climate emergency by writing a letter to a public official or participating in a rally, a walk out, or a strike. (This is an extraordinary participation percentage.) Clearly, our children are concerned, and they should be. Our job as parents is to help them understand that through actions, their angst, their grief, and their fears can be channeled to build active hope.

It is important to remind them that in a country as big and diverse as the United States, and in many countries around the world, people are still free to march, to express personal opinions, and to have their voices heard, even when our personal ideas don't jibe with the opinions of those in charge. Our grown children are never too old to be reminded how, as a nation, we have fought for generations and over decades for these rights—rights that are not enshrined in many countries around the world. However, even in the United States of America, these rights are regularly and increasingly being threatened. As I write this book, conservative lawmakers have put forward laws *criminalizing protests* in at least 18 states since 2017; civil liberties advocates say these laws are unconstitutional.[3] Vested interests, including those with ties to the fossil fuel industry, are working in many states as well as nationally to pass laws that will limit public protest and free speech and silence protest votes—restricting, and in some cases actually banning, protests of all kinds.

This should shake us to the core. We, and our grown children must take every opportunity we can to vocally and visibly protect the laws—put in place over decades and fought for over centuries—that protect our right to speak out, to gather, to be seen and to be heard. Our rights are much more fragile then we realize, so it is critical to protect them as we work together with our children to create a livable present and future.

Developing youth leadership as a parent requires that you provide logistical support and model ethical behavior in your dealings with others. It also involves understanding that youth leaders are good for society because they are the best equipped amongst humanity to make choices for the collective best interest. Our youth can make these kinds of decisions because young people are fairly unencumbered by immediate pressures of obligation and responsibility. And, finally, it means relinquishing power to young people. My supporting role in Anna's work feels like a very, very long unpaid internship. However, there are massive intrinsic rewards attached to helping stabilize the climate for my daughter and her children, and for your kid. I want thousands of leaders like Anna in the world—this is what has become necessary in the climate crisis. One climate organization touts "We Need Everyone, Everywhere." Well said. We need to collectively figure out how to coach and support a thousand other youth leaders-in-waiting, keep their egos in check, as they develop strategies for actions.

—Kathy Mohr-Almeida

At the same time that we fully embrace and support youth climate activism, their actions must renew and enhance our own commitments, not replace them. Young voices are bringing much-needed attention and focus to our climate emergency, but the responsibility to take action remains with us all. At the 2017 Women's March in Washington, DC, three generations of my family marched: my 83-year-old mother-in-law, my 16-year-old daughter, and my 12-year-old niece. In the seemingly endless sea of hundreds of thousands of people, I looked to the left, to the right, behind me, and in front of me, and I saw determination to make change mirrored on the faces around me. When I got home, I looked in the mirror and said to myself—out loud—*Look no further—it's you*. While I knew this, and while I regularly remind others of this, I was not fully and consciously putting this mantra into practice. We all have different abilities, opportunities, and powers with which to make a difference. For me, the Women's March put a spotlight on this reality. We can't rely on someone else to fix what we think needs fixing. We, therefore, must not wait to be asked, but must take it upon ourselves to find ways to work collectively to fix what we see is broken. Our youth are being heard around the world; we must show them we are working at home to put their demands into action.

There are so many marches, rallies, and demonstrations that take place every day, across the US and around the world—marches for jobs, for teachers, for equal pay, for tax reform, for healthcare, for immigrant rights, for women's rights, for science—and many more issues. Some of these are reported in the media, but most are not. All are opportunities and outlets for people to stand up for what they believe and to demand that our elected officials pay attention and listen to what we have to say and what we care about. Being in the streets of your city, your town, your campus, or your neighborhood, with others who share your passion and your concerns, is incredibly empowering. Being there with your children at your side is doubly so. Marching and making our voices heard is a privilege and right to be protected at all costs. Yet, does the world change direction because 400,000 people marched for climate action in the streets of New York City in 2014, because millions of pink pussy hats took over the streets and cities of the United States in 2017, or because high school students demanded to be seen and heard in 2018 as they voiced their concerns and their right for safety in their schools? Maybe not. But with increasing frequency, our youth are taking a lead and speaking truth to power on these critical and interwoven issues.

In 2018, young people across the world began climate striking. This student-led movement includes Fridays for Future and School Strikes for Climate. These strikes grow each week, as children from grade school to college demand action and that we, as parents and adult allies, give our attention to the urgency of our climate crisis in direct, visible, and pervasive ways. This youth-led activism had its spark from a single 15-year-old, standing alone, for the first time, in front of the Swedish parliament buildings in the summer of 2018. Greta Thunberg's demands were simple, that there be *Fridays for children in the future*. Greta was demanding that the Swedish government wake up to the urgency of the climate crisis and take action. Her voice has now been heard around the world as millions of children have joined her and as parents of children too young to strike on their own and parents supporting their striking children have been activated alongside their children, worldwide.

I wish we, as a society would get our act together as quickly as possible, so those students don't have to sacrifice their education anymore.

—Helene Costa de Beauregard

Our children are growing up in a world where partisan politics are pulling us as a nation, and often as families, into many divisive parts. Yet our youth—our children, in middle, high school, and beyond (and sometimes even younger)—are demanding we find ways to come back together to work collectively on gun violence, immigration, equity, gender equality, voting access, and other issues of acute concern—issue areas that are all made worse by our climate crisis. In each of these categories, you can point to young people (and groups started by youths), who are leading and truly making a difference and changing our world for the better. An exciting youth-led and youth-created umbrella organization that brings together youth leaders and youth organizations representing justice, climate, voting rights, and more is the Future Coalition.[4]

Remind your grown children what the Constitution says: "We the People... in order to form a more perfect union, establish justice, insure domestic tranquility, provide for the common defense, promote the general welfare, and secure the blessings of liberty to ourselves and our posterity." These intertwined concepts were identified as vital areas requiring protection long before we faced our climate emergency; we must not let them be siloed now. Together they are part of how we will ensure a just transition and a more stable climate. Addressing voting rights; protecting our democracy, equity, and equality; and ensuring equal access to clean air, healthy food, clean water, and a safe environment are all part of the solution. Focusing on one part helps the whole, and understanding how these parts all fit together is critical.

I left a 25-year career in high tech six years ago to dedicate myself to working on climate change; this work is pretty ever-present in our household. My kids know I'm working on it, both locally and around the country. They've both had the chance to get involved from time to time in some of the work as well. The latest developments in the climate movement, from Trump backing us out of the Paris Climate Agreement to the Sunrise Movement and the Green New Deal are dinner table topics.

—Larry Kraft

Climate action and working on solutions can and must be intergenerational. We all need to take the lead by doing what we can. Finding opportunities to work with your adult children, and perhaps with your grandchildren, reinforces your family's commitment to our collective future and the future of those closest to you.

The best part of being active in the environmental and climate justice movement is that it's intergeneration. My girls have grown up seeing people of all ages showing up to fight for justice and for climate action. They know that community meetings, organizing protests, growing our own food, composting, testifying at hearings, holding polluters accountable, and getting others involved are all part of the climate solutions. More importantly, they have a role to play too—not just observing the action or studying it—but being a real part of it. They have experienced the joy and love of community-led action and activism, and I hope that these early experiences grow in them a sense of their responsibility to be awake and engaged in the struggles for climate justice.

—Ana Baptista

7

Creating a Million Ripples

From Me to We

The first time that we, as a species, viewed a photograph of our home was in 1968, when a picture taken by astronauts on the Apollo 8 space mission was sent back to NASA. Millions of viewers were glued to their televisions, and together we went from *me* to *we*—we saw our planet rising above the moon's surface, a pale blue dot set in one frame, suspended in space. It seems that a collective sigh was released around the globe that day. An understanding, recognition, and realization were born—our planet is fragile. And we began to realize that we would need to find ways to come together to protect her. Within two years of the 1968 Earth Rise photograph, we celebrated the first Earth Day. In the US, a Republican president established the Environmental Protection Agency, and the first national Clean Air and Clean Water Acts were established.

Today, with hundreds of channels and a seemingly endless stream of media sources, it is increasingly uncommon for millions of viewers to watch the same thing, at the same time, around the world. More than 50 years after our first glimpse together of our home, our planet is healthier in some respects than she was then, and yet infinitely and precariously worse.

The Earth—our home—has been around for billions of years and will continue being around for eons more. The question we must ask as we enter the epoch of the Anthropocene is, "Will we?"

We stand now where two roads diverge.
But unlike the roads in Robert Frost's familiar poem,
they are not equally fair. The road we have long been traveling
is deceptively easy, a smooth superhighway on which we progress
with great speed, but at its end lies disaster. The other fork of
the road—the one less traveled by—offers our last,
our only chance to reach a destination that
assures the preservation of the earth.

—RACHEL CARSON, writer, scientist, ecologist,
environmentalist, marine biologist, *Silent Spring*, 1962

The good news from my Climate Mama vantage point is that so many more people are engaged, aware, and taking action. And they are demanding that those who can go big to protect our species—and all species—do so with the urgency that our crisis demands. People all over the world are finding ways to be more thoughtful and deliberate with actions that will impact the planet, at the same time as more and more are waking up to the emergency we face.

Be inspired. Help is here.

Those of us who have been working on climate issues for years experience a multitude of feelings as we watch the exciting impacts of new climate activists—young and old—on the scene. These people are digging in and taking action. For me, the first thing I feel is a huge sense of relief; this existential crisis isn't resting squarely on my own shoulders. I can take a deep breath—maybe even some time off—as others join en masse to carry the burden and the knowledge forward. I am truly confident that we can and are creating lasting and active hope. I also stand aside sometimes, as I listen and watch with bemusement, as new friends and colleagues rightly and indignantly state, "If only others would wake up and take action immediately, we would solve this so much faster!"

I get it, and I so agree. I remind myself regularly to say thank you to all, acknowledging those who have come before, those who remain, and those who are just jumping in. From my perspective, reminding our children through examples of kindness, even as we face our urgent reality, matters more than ever. We mustn't forget, in our haste to take climate action, that the lessons of compassion, kindness, justice, empathy, and thoughtfulness, particularly as the world seems to spin out of control, remain our parental responsibility. These lessons are key to building resiliency and strength so that our children remain strong and directed—so that they weather the storms ahead and become leaders themselves, inspiring others through their actions, deeds, and words.

Are you a fan of the *Avengers* movie series? Spoiler alert, in case you haven't seen *Avengers: Endgame*, stop reading and skip to the next page.

Imagine for a moment that we can travel back in time, keeping all our knowledge from our current time period. In *Avengers: Endgame*, the characters invent a time machine to travel back to find the infinity stones that will bring back half of all life in the universe (those who were disappeared by Thanos in his twisted quest to save the universe from the damages our species brought upon itself).

In a similar way, with a less dramatic wave of the infinity stones, let's for a moment travel back together to the *Titanic*: we are on board the day before it crashes into the iceberg that sank it. We know what's going to happen. Yet everyone around us is unaware; they are oblivious to what tomorrow will bring, simply enjoying the moment they are in. Some are in awe of the beauty around them, admiring nature's incredible ice sculptures; others are simply mesmerized by the ship itself. Yet, we know what's coming and no one believes us. No one hears our warning cries. Our children have alerted us, and we have shared their news with the ship's crew. Our children's eyesight is strong and clear. They see the iceberg and know we are on a crash course. But this is 1912. Children are to be seen and not heard. Passengers and crew alike aren't sure what to do about us or our children. Some ignore us; others suggest we seek professional help; and others try to talk sense into us. It's common knowledge that the *Titanic* can't turn on a dime. But we know that if the process of changing course isn't begun immediately, we will hit the iceberg and many people will lose their lives. What do we do? How do we make people see, hear, and believe us and our children?

Clearly and frustratingly, it shouldn't be so hard for us to break through the chatter of ignorance. Scientists are shouting from the rafters, ringing alarm bells. Mother Nature is working overtime to wake us up. Children's voices and faces have replaced polar bears as the call to action.

Thinking in Systems and Channeling Systemic Change

Change can happen in the blink of an eye, roaring in like a tsunami, unexpected and with momentous force. Yet in most instances, the tsunami has been built from a million ripples.

In 2006, my son and nephew were ring bearers at my brother Michael's wedding to his husband, Phillip. In 2005, Canada became one of the first countries to legalize same-sex marriage. Marriage equality was a long time coming—built on hard work; the letting go of preconceived, outdated, and prejudicial beliefs; years of meetings; protests both violent and peaceful—all parts of the same movement to foster equality and champion love. Marriage equality isn't universally accepted, but it is a growing and recognized right in many places around the world. Sadly, seemingly inalienable rights, like access to clean water, clean air, a livable future, education for all, and freedom of speech, remain unsettled, unclear, or inaccessible for too many people and in too many places across the globe.

Our climate movement is growing in jumps and starts, but real transformational change is beginning to break through.

And yes, I know we need a system change rather than individual change. But you cannot have one without the other.

—Greta Thunberg, Brilliant Minds Conference, Stockholm, 2019

Systemic change is built on hard work that often goes unrecognized and unacknowledged. When systemic change happens, it sometimes seems like magic, like that moment was *meant to be*, transformative in its place and time in history. Women's voting rights, reproductive rights, anti-slavery laws, and LGBTQ rights are all systemic changes that have been made, but they all occurred because people followed a path that had been put in place by others. Sometimes the path is clear, meticulously marked and easy to follow. Other times its worn away; trail markers are broken and scattered across decades, and even centuries. But the path is there nonetheless, regularly cleared, sometimes by individuals acting alone and other times by groups working together.

Unfortunately, many of these hard-fought-for rights—seemingly safely ensconced in law—are being repeatedly challenged today. We can never truly let our guard down. In places around the globe, many of the rights we consider settled are not recognized or codified into law. In 2019, my daughter's right to control what happens to her own body, something fought for when I was a child, something that seemed settled and protected in the US decades ago, is once again under new (and increasingly aggressive) threats. We need to be watchful and constantly on guard.

Scaling Solutions

Professor Erica Chenoweth,[1] in her research on non-violent direct action, tells us that to create systems change, we need to mobilize 3.5% of the population. What would 3.5% of the people you work with look like? Of the parents in your children's Parent Teacher Organization or of people demanding change from your state legislature or from Congress? These numbers seem very doable. Together, we can create the change we want, change we can see, and the change that will allow our children to thrive.

Let me share a personal story about being arrested at the 2011 Washington, DC, Tar Sands protest. A decision to risk arrest should and must be made carefully, with a clear idea of what can happen and, whenever possible, with ongoing support and help from others who have had significant organizing experience. In my case, I made this decision not only because it was a direct way I could bring attention to the climate crisis but also because I could. I was coming from a place of privilege: I am a white, middle-aged woman with no previous arrest record; I have the ability to take a day off work without penalty from my employer or the need to worry because of lost income. I was well prepared and informed about what would happen, and there were people to support me every step of the way. I could do this for so many reasons, and I had the opportunity to represent so many others who wanted to be there but could not. I was not alone.

At our mandatory civil disobedience training the night before my arrest, we were instructed not to bring anything to the protest that we could not afford to lose, including cell phones and wedding rings. (Do you know how hard it is to get a wedding ring off your finger, after it has been in place for 15 years?) We were also told to dress in a dignified style, as if we were going to a business meeting. I had packed a skirt and top, neither of which had pockets, and I hadn't thought through how I would carry my ID and $100, the two items we were told to take to jail. The ID was so we could be properly booked, and the $100 would be our *get out of jail* card. The hope and best-case scenario was that whatever the charge was, we would be given the opportunity to *post and forfeit*, allowing us to pay a $100 fine and leave jail the same day.

This is where a bra comes in handy. That morning as I dressed, I put my driver's license, $100, and a metro card into my bra, which effectively served as my purse for the day. I did move these items to the waistband of my skirt just prior to my arrest and before I was handcuffed. I hoped this would make it easier for me and for the arresting officer who would frisk me. Yes, I was frisked, in fact several times, and yes I was put in handcuffs—the plastic strap kind. These stayed on me from the moment of my arrest until I was processed at the jail. That day, I also rode through the streets of Washington, DC, in a police van with bars on the windows and a police motorcycle escort as I watched the White House fade into the distance. Many firsts for me that day.

We can do this.

Yes, it would be great if everyone joined us, but we don't need, nor will we realistically be able to mobilize, everyone. Remember that 3.5% figure. It is doable. Being part of a movement that you believe in builds strength, hope, and community. Our children are watching us; their present and future is in our hands.

We will do this.

We must also stick to our demands that those who can go big do so; we will not be satisfied with half measures. Utility, gas, and oil companies all need to become renewable energy companies—or they need to shut down. We need multinational corporations of all kinds to operate with a zero emission, closed-loop mentality. And we need everyone to consider the fragility of our global commons—our air, water, land, and sea—when making corporate and personal decisions. Our planet has shown us she has boundaries; we are exceeding them. We need all elected leaders to be working on climate action plans and on the implementation of their plans. These plans must be comprehensive and inclusive, ensuring a transition that incorporates the needs of all people. Clearly this will be hard work, but transitions and transformations are coming, whether we want them to or not. We still can navigate the changes to come, but our time at the wheel is running out.

Inertia and fear must not slow us down. We also must recognize that we don't have to wait for someone else to lead. There is no order or linear progression that is the right road forward. Moving forward is the way to go. For some of us, this will be by leaps and bounds, and for others, one step at time. Each of us must do the most that we can as we live our lives in the Anthropocene.

We therefore need to begin imagining what the future we want looks like. We need to talk about this future and regularly share our vision for what it will look like with our children, friends, family, and beyond. This will be a vision of something new, something different, something hopeful, and yet something concrete—something we can build and create together.

You never change things by fighting the existing reality.
To change something, build a new model
that makes the existing model obsolete.

—Buckminster Fuller

In 2007, shortly after my first climate reality training, Al Gore announced a 10-year plan, how to get to 100% renewable energy, create good jobs, and advance climate solutions. Sound familiar? In 2019, with research and support from the Sunrise Movement, Congresswoman Alexandria Ocasio-Cortez and Senator Ed Markey launched the Green New Deal resolution in both the House and the Senate. The Green New Deal is a 10-year plan to build a better, safer, and livable tomorrow, based on the following five broad goals:

1. Achieve net zero greenhouse gases by 2050.
2. Create millions of good, high-wage jobs, ensuring prosperity and economic security for all people of the United States.
3. Invest in the infrastructure and industry of the United States to sustainably meet the challenges of the 21st century.
4. Secure clean air and water, climate and community resilience, healthy food, access to nature, and a sustainable environment for all.
5. Promote justice and equity by stopping current, preventing future, and repairing historic oppression of frontline and vulnerable communities.

The details of the plan were purposefully elusive for the 2019 launch, but millions of people around the country and the world immediately began discussing the plan, and tens of thousands began envisioning what the details can and must be. Green New Deals at the state and local level are popping up with regularity. Everything old is new again; everything has its time. As we teach our children, *if at first you don't succeed, try, try again.*

Being the Change We Wish to See

I hear it all the time from dear friends and family and often from complete strangers, too: *just give me three things I can do that will help.*

Life is messy, and the world is a very messed-up place. My best and most honest advice is to remind your children at every step of their lives to lead with their passions.

I feel strongly that if we all harness our determination for positive change, and act on what we are passionate about, we can make the world a sustainable place for us, for our families, for our children, and for our collective home. Our journey will be long, it will take the rest of our lives. It won't take everybody, and we don't have to convince everyone. We just need to keep moving forward, one step at a time, no looking back. When people ask, "What's next?" I tell them it's messy and complicated and will not be easy. We are fighting for our lives and a livable future for our children.

As Martin Luther King so eloquently reminded us, "The arc of the moral universe is long but bends toward justice."

For our journey to be truly successful, we must find ways to stand together for climate justice, human rights, and equality, for love and against hate. We must stand together for the full respect of science and for immediate and long-term actions that build solutions to the global crisis of climate change. That being said, having three things to rally around can certainly be a good conversation starter. So, in helping you build your elevator pitch for why and how you are acting on solutions to the climate crisis, I will share my top three things that you can share with others. Each of these ideas is scalable. You can ride your passions as far as you let them take you by doing the following:

1. Involve yourself in our government systems. Vote, consider running for office, or find other ways to be a voice for democracy.
2. Follow up and be prepared.
3. Plant seeds and nurture them.

Vote, Vote, Vote

Whenever you have the opportunity to vote in an election, on any issue and however insignificant it may seem, do so. Voting rights (and, by extension, any participatory democracy) are usually hard fought for and hard earned. We must cherish, foster, and protect these rights. Throughout history, wars have been won and lost and countless lives have been sacrificed in pursuit of the right to vote and for free and open democracies. Currently, it surely seems that our US democracy is being stretched and bent. (Perhaps to its limit?) We must not allow it to be broken. As we face more and more challenges from our climate crisis, democratic systems around the world will be threatened to their cores. Understanding this, we must be on guard. We need to do what we can to ensure and encourage thoughtful candidates who have clearly defined climate action plans to run for office at every level of government. And we can and must work on getting people of all ages and from every neighborhood to vote in every election. As voters, we must do our homework and make sure that the candidate we choose has a ready-to-implement and well-developed climate action plan.

When casting your vote, do so with the intent of showing your children, your country, and the world that you completely and utterly repudiate hate and lies and that you stand up for science and for the truth. There is no room for complacency, and no time to prop up elected officials who do not take the crisis at hand with the urgency of action that is required. It is often hard to connect the deliberate and singularly focused act of voting with the urgency of our times. But you must remain conscious of the importance of each and every vote, even if it seems like only one small step. Do not allow yourself to be stuck, frozen, or paralyzed, or to feel so much disappointment or disgust with the system and those running it that you do not vote. Remind your children of this as well. They may seem young and far away from voting age, but the importance of casting a vote is developed at an early age. Since the beginning of the 21st century, young people are no longer voting in high numbers in the United States and in many other countries around the world. This must change; the present and future requires it.

Consider running for office, and encourage people you think would be good leaders to do so, too. Remind your children that when they are grown, they may want to run for elected office as well. Gandhi's words ring loudly: "Be the change." Each level of government—local, regional, state, and national—has a role to play in creating the polices, rules, and regulations that can help us respond to our climate crisis. At the same time, if people who dismiss or deny the urgency of our crisis run for office and win, they can slow things down to a point where our future can and will be catastrophic. By voting, and by supporting people who are willing and able to make hard decisions based on a clear understanding of science—hope and change can find footings to not only survive but to thrive. There are no sure answers, but if we envision a system that can and will react to the harsh and difficult realities we face, we can create it, nurture it, and watch it grow.

In 2009, I was appointed to my city's Green Living Advisory Board and eventually became its chairperson. The objective of this board was to educate our city residents on how to live a greener lifestyle.

I had the privilege of lobbying on Capitol Hill in Washington, DC, for federal climate change legislation. This eye-opening experience made me realize that the only way to effect the biggest changes was to get elected officials to understand the crisis and do something about it.

With the Washington, DC experience under my belt, I realized I needed a louder voice to address the escalation of the climate crisis in my hometown. So, I decided to run for city council. After six months of campaigning, in 2018 I was elected to the Boca Raton City Council, where I now create policy and direction to manage the city.

My hope is that my children see my commitment to our environment and follow in my footsteps as they become leaders in their own communities.

—Monica Mayotte

Follow Up: Preparing for our Climate Emergency

- **Step one:** Vote for and elect candidates who promise to show up each and every day to build resilient communities and work diligently and openly on climate solutions.
- **Step two:** Follow up with them regularly.

We need elected representatives to follow up on their promises. It's your job, as concerned and active citizens, to watch them, talk to them, and demand that the plans they have promised and created for addressing the climate crisis are put into action. Use these opportunities to teach your children the importance of following up—following up on their own commitments as children, young adults, and, eventually, as adults. As well, teach them the importance of follow-up from those we entrust to follow up for us: elected officials and others who make commitments on behalf of a community or group. Did they see their promises through? Take your kids with you to your elected officials' offices. Have your children write letters and cards to these officials. They can draw pictures if they can't write—perhaps they could draw a picture of their favorite things that climate change could impact if we don't take action now.

Having a plan at work, at school, and in public places for emergency situations like earthquakes, active shooters, and fires is common. Emergency drills are done so people will know what to do when an emergency strikes. Ensuring that your community is prepared for climate emergencies should also be second nature. Climate drills should become as commonplace as fire drills. If you live near a river, an ocean, or in a flood zone, a lifeboat is usually within reach. We must make sure our children's schools are equipped with lifeboats of all kinds. Demand that your school district practices climate emergency drills as part of extreme weather planning. This will help educate your community about our climate emergency and protect and prepare your children as well. Do the same at home.

The 2018 Paradise, California, fires tragically demonstrated that extreme events happen quickly—and sometimes with little or no warning. Being prepared and understanding the risks that the climate crisis presents to your community is critically important. Individual gas masks are a requirement for each school student in Israel, and air raid drills are a part of each Israeli child's routine. Similarly, all schools located near forests and forest buffer zones in fire-prone areas should now be equipped with breathing masks for each child. Not only must we follow up with our own family plans, we need to make sure that our elected officials and the companies we choose to support follow up for our broader collective as well.

Plant Seeds: Literally and Figuratively

Literally and figuratively, we all must plant seeds. Scientists are telling us that a managed, targeted tree-planting campaign—around the globe—can mitigate and slow down our climate crisis.[2] We must reforest our planet so that Mother Earth can help do the heavy lifting by drawing down excess carbon and begin her recovery process. Tree planting is an activity you can do as a family; encourage other families and friends to join you. Planting native species gardens, rain gardens, growing food locally in our parks and back yards, and participating in community gardening are all ways to help your children get their hands dirty and connect directly to Mother Earth. In our 21st century, children too often grow up far from nature's magic. Reminding our children of where our food comes from, how the planet regulates and cares for the millions of species that have evolved, and the interconnectedness of all of these critical parts are important lessons to share with your children.

Figuratively speaking, we must help plant the seeds for our future. As we demand that those who can go big do so, we can and must continue to do the smaller things to show we care and that we are serious about taking climate action. We must make this clear in everything we do.

From Me to We

Over the course of my life and throughout my career, I have had the pleasure and the honor of meeting and spending time with presidents and prime ministers, kings and queens, governors, scholars, religious leaders, renowned scientists, and CEOs. I've had lunch with ambassadors and dinners with billionaires; I've gone on archeological digs with world-renowned paleontologists and conversed with international leaders in business and science working on transformative and revolutionary new approaches to prepare, train, and build our 21st-century world.

I've also had the pleasure and the honor of standing side by side with mothers and fathers from farms, rural communities, and big cities. We have held hands and held our ground both on the front lines of fossil fuel infrastructure expansion and at the doors of city halls, state houses, Congress, and the White House. Who is braver, stronger, or more powerful: the parents or the dignitaries? In different situations, each one of us can and does play the most impactful and important of roles.

Our history and our present is rife with visionaries and leaders who have sparked movements, created and invented world-changing discoveries, and ignited and built active hope. Martin Luther King, Rachel Carson, Rosa Parks, Elon Musk, and now Greta Thunberg are but a few examples. Who will set in place the systemic changes and giant leaps that our climate crisis demands? Will it be your child? While we wait anxiously for those giant leaps, each one of us can create ripples; and together, our millions and millions of ripples can turn into a tidal wave of real and sustainable hope and action. By visioning this, we can create it.

What I have found after spending more than 25 years working in and around the climate crisis is that there are many positive, enlightened, and hopeful people and concrete solutions already in motion and working. The good news perhaps, and the reality that can be frustrating at times, is that no one has the definitive answer that will solve the crisis at hand.

We must become comfortable with uncomfortable truths. There is no easy fix. We don't know who should determine *the* path forward. No one can tell us with certainty that if only we did this first, our planet would heal and humans would have a very real chance at survival.

The ultimate path may well be found by our children or their children. Children have a way of seeing through to the truth and aren't blinded by social norms, beliefs, or platitudes. But in the meantime, we must be very careful not to transfer the weight of the problems we face from our shoulders to those of our children.

My hope is that this book has helped you understand this. There are no silver bullets or magic seeds when it comes to climate solutions. We can't wait for those to arrive, nor should we put our hopes in them. They will not be coming. They do not exist. The sooner we face this hard truth, the better.

There is an old adage that says wisdom comes with age. I am positive that this is not always the case. Until recently, most of us taught our children to respect their elders and put trust in their wise counsel. Historically, we have looked to those we elected—to the women and men we entrusted with our votes—to guide us, to represent us all once elected, and to do what is right for our country, our legacy, land, and future. Clearly, this is no longer the case; whether it ever was remains something to ponder. Too often, those with power use it for their own personal gain rather than for the betterment of us all.

What we do know, particularly when it comes to the climate crisis, is that there are many individuals, companies, and organizations fighting *against* the truth, consistently sowing doubt about the reality of our climate emergency and working actively to slow down the urgent responses required. We must stand strong against the lies and the false belief that we have infinite time, or even a long time; we must push back loudly and hard. Our children are counting on us; they are watching us.

Our hopes, our opportunities, our optimism, and our reality must come from the solutions that do exist and which we each can champion, either independently or together. My enduring hope and what keeps me going is my firm belief that each of us has a role to play in building and creating our livable tomorrow and today. When we get tired, we can and we must allow ourselves to take a break, to know that others have our back and will replace us on the front lines. The future doesn't rest on our individual shoulders. Rather, our present and future and that of our children is carried on the backs and in the arms of each one of us, together.

I have found in the midst of the sadness, the grief, and the loss that climate change brings that taking time and finding ways to connect with what we love can renew our passions and our hopes. Spend time with those people in your life who make you laugh, who bring you joy, and who truly love you and whom you truly love. Take time to cherish and be with and in nature. Immersing ourselves, our children, and our families in our natural world is both profound and spiritual. Walk in the majesty and the shadows of mountains rugged and wild; watch a tiny ant carry a stick that is twice as large as she is; these are acts both healing and life affirming. Nature is wonder, and wonder is found everywhere nature is—in an abandoned lot in the middle of a city where a flower has appeared in concrete cracks, or out in the wide open African plains, overflowing with a multitude of species—we need only open our eyes wide—nature is everywhere.

One final paragraph of advice: do not burn yourselves out.... It is not enough to fight for the land; you also need to enjoy it. While you can. While it's still here. So get out there and hunt and fish and mess around with your friends, ramble out yonder and explore the forests, climb the mountains, bag the peaks, run the rivers, breathe deep of that yet sweet and lucid air, sit quietly for a while and contemplate the precious stillness, the lovely, mysterious, and awesome space. Enjoy yourselves, keep your brain in your head and your head firmly attached to the body, the body active and alive, and I promise you this much; I promise you this one sweet victory over our enemies, over those desk-bound men and women with their hearts in a safe deposit box, and their eyes hypnotized by desk calculators. I promise you this: You will outlive the bastards.

—Edward Abbey[3]

I have come to the realization and acceptance—through life experience, through the passage of time, and through the joys of birth and the sorrows of death—that our lives truly are fleeting. This fact remains both vividly transparent and at the same time deeply hidden to most of us. Knowing and understanding that, in reality, our walk on this planet is but a blink of an eye makes it doubly hard to accept and acknowledge that we are living in and have helped create, the Anthropocene. We are now at a point where those of us alive today, at the start of the 21st century, have the ability in very real and concrete ways to impact the future existence of our entire species and of many other species alive today. This fact is truly mind blowing; it is weighty and complicated and also incredibly challenging and burdensome to get our arms and minds around. Yet, we must.

One of my primary goals in writing this book is my enduring hope that it will fortify and strengthen your hopes and your resolve. Our shared belief in a positive future, coupled with constant and forward motion, will sustain us as we build a livable present and future for our children and for us. We must trust in one another, in our indomitable will to preserve our species as we claim our survival. We must be open to and accept that transformations will happen. Changes will occur—changes that we currently may not yet be able to truly visualize. We must trust in these changes; they are required.

We must also visualize the future we want, confident and optimistic that by helping this future unfold we will ensure that our children can and will live their fullest lives We must approach this with exuberance and with active, intrinsic, and radical hope, as well as with positive and directed purpose actualized through our actions and our voices. As 15-year-old Autumn Peltier told the United Nations in September 2019, "One day, I will be an ancestor and I want my descendants to know I used my voice so that they could have a future."[4]

As this book started, so it will end:

We will be coming to terms with the urgent realities of our climate crisis for the rest of our lives.

We all need to get comfortable with this fact—in a very uncomfortable way.

With sincere love and thanks,
Your Climate Mama

Resources

My list of trusted resources on climate education, science, justice, and action expands regularly. I could easily fill this entire book with credible resources on our climate emergency. I am therefore limiting each heading to "10 picks," listed in no particular order, in the following seven categories: Youth-Led or Youth-Focused Organizations, Parent-Focused Organizations, Climate Action and Educational Websites, Climate News Websites, Climate Justice, Notable Books, and Climate Science.

The ClimateMama.com website contains many additional resources, blog posts, and profiles of Climate Mamas and Papas from around the world who are leading on our climate crisis and showing us how to take action and lead too. Be sure to check it out!

Youth-Led or Youth-Focused Organizations

Sunrise Movement
Zero Hour
Future Coalition
Imatter Youth
Earth Guardians
Earth Uprising
Earth Charter Indiana
Young Voices from the Planet
Fridays for Future
Extinction Rebellion Youth

Parent-Focused Organizations

Mothers Out Front
Our Kids' Climate

Climate Parents
Moms Clean Air Force
Parents for Future (UK, Germany, US, France, and more)
Parents Roar/Föräldravrålet
Mothers Rise Up
Sachamama
Dear Tomorrow
Utah Moms for Clean Air

Climate Action and Education Websites

Our Children's Trust (US)
The CLEO Institute
Climate Reality Project
Alliance for Climate Education
Schools for Climate Action
Citizens' Climate Lobby
The Years Project
Human Impacts Institute
Project Drawdown
National Wildlife Federation—Climate Classroom Kids

Climate News Websites

Inside Climate News
Climate Nexus
Climate Crock, Peter Sinclair
Skeptical Science
The Guardian—Environment & Climate
350.org
Climate Central
DeSmogBlog
Photo Ark: National Geographic
Yale Climate Connections

Climate Justice

A Just Climate.org
WeACT for Environmental Justice
Women's Earth and Climate Action Network

Climate Justice Now. Earth; Jill MacIntyre Witt
Earth Justice
Laudato Si: On Care for our Common Home; Pope Francis' Encyclical
Mary Robinson Foundation—Climate Justice
Climate Justice Alliance, Communities United for A Just Transition
Poor People's Campaign
Indigenous Environmental Network

Notable Books

Active Hope, Joanna Macy and Chris Johnstone
Climate Generation: Awakening to Our Children's Future, Lorna Gold
Uprisings for the Earth: Reconnecting Culture with Nature, Osprey Orielle Lake
Moral Ground: Ethical Action for a Planet in Peril, Kathleen Dean Moore and Michael P. Nelson
The Weather Makers: How Man Is Changing the Climate and What It Means for Life on Earth, Tim Flannery
An Inconvenient Truth, Al Gore
An Inconvenient Sequel: Truth to Power, Al Gore
Drawdown: The Most Comprehensive Plan Ever Proposed to Reverse Global Warming, Paul Hawken
The Parents' Guide to Climate Revolutions, Mary DeMocker
Storms *of My Grandchildren: The Truth About the Coming Climate Catastrophe and Our Last Chance to Save Humanity*, James Hansen

Climate Science

National Academies of Science
Real Climate
NOAA Climate (National Oceanic and Atmospheric Administration)
NASA Global Climate Change
NASA Climate Kids
Tyndall Centre for Climate Change Research
The Intergovernmental Panel on Climate Change (IPCC)
Potsdam Institute for Climate Impact Research
Nature Climate Change
American Meteorological Society's *Journal of Climate*

Climate Action Organizing and Book Club Guide

Are you beginning to meet with friends, or thinking about getting together specifically to discuss your growing concerns around our climate crisis? The following discussion guide will work both for an established book club discussing this book and/or for a broad discussion with concerned friends and colleagues interested in coming together to plan, to discuss, and to share ideas on ways to organize both collectively and as a family around our climate emergency.

Broad Topics for Discussion

1. How does this book make you feel? Hopeful, determined, anxious, sad? Can you relate easily to what is being discussed throughout the book, or are the sentiments expressed difficult to reconcile with your current thinking? And if so, why?
2. Are most things in this book new news to you? Share where you are in your journey. Are you just starting out and trying to figure out how to become more involved, or are you already active with a climate-focused organization or group?

After sharing your thoughts about the book broadly, have a focused conversation on one of the three following questions. If time allows, consider expanding the discussion to a second or third question. Agree to convene at a time in the near future with participants who are particularly engaged, so that you can keep the conversation going.

Focused Discussion and Possible Actions to Take

1. How is our climate crisis impacting you, your family, your community? Discuss what you see, hear, feel, and understand.
   Action: Create an "elevator pitch" to explain why you are concerned about the climate crisis. You should be able to deliver it in three minutes or less. Begin by writing out your story—not necessarily climate related. (Climate Generation[1] has a great story-telling guide, which may help with this.) Make your elevator pitch something that even your partner's Uncle Bob will be moved by. Break into small groups or even just groups of two, and practice together; share your story, and have your partner share hers.
2. What is your vision for the future? Take time to consider it, write it down, and then share it with the group.
   Action: Create a personal climate pledge. On one page of paper, highlight your vision for the future 5 years, 10 years, 20, and 30 years down the road. Begin by writing down your hopes for your children and your family; what do you see them doing? Where will they be in 5, 10, 20, and 30 years? Who will they be with? What will they be doing? On a second page, create an action pledge for your family to help actualize your vision, particularly in the near term. Date the pledge, sign it, and agree to revisit it together as a group six months from the date you signed it.
3. Acknowledge and discuss the immensity of the challenges at hand. Have an open discussion about how this makes you feel. What stage of grief are you at? How are you dealing with it? How are you moving from resolve, to hope, to action? Chapter 3 can be of help with these issues.
   Action: Create a support group from members of your Book Club for this exercise, but consider inviting others outside of the Book Club to join you as a member of your personal climate support group. Meet on an ongoing and regular basis. Assign partners for each member of the group. Each person should have a person who, when fears or sadness become too much to handle, they can share their feelings with. Or conversely, when you experience a

personal climate win, create a family solution, or participate in a climate action, you can share this with your partner.
Action: Identify a local climate action group, and agree to go together to their next meeting or action.

4. What questions are your children asking you about our climate crisis? What advice, experiences, examples can you add to the conversation on ways to answer your children and engage them in age appropriate action?
Action: Share questions you are being asked by your children. Create a family, multi-family, or community climate action plan with your children. Follow up at regular intervals with family meetings to discuss how it is going and how your family or group is moving it forward.

E-mail questions, thoughts and ideas to: info@climatemama.com.

Notes

Dedication

1. This letter to my children was first published May 5, 2016 on Dear Tomorrow.org, a digital and archive project established for people to personally connect with the issue of climate change, to commit to taking stronger action, and to share these stories with friends, family, and their social networks.

Preface

1. Under President Obama, the United States saw the largest expansion of oil production in the history of the country, including an explosion of fracking and gas developments. See "The Irony of President Obama's Legacy" by Robert Rapier, January 15, 2016, *Forbes*, forbes.com.

Chapter 1

1. Harriet Shugarman, excerpt from *Global Chorus: 365 Voices on The Future of The Planet*, edited by Todd E. Maclean, Rocky Mountain Books, 2014.
2. David Wallace-Wells, *The Uninhabitable Earth: Life After Warming*, Tim Duggan Books, February 2019.
3. World Resources Institute, "New Global CO_2 Emissions Numbers Are In. They're Not Good," December 2018, wri.org/blog/.
4. Elizabeth Kolbert, "Enter the Anthropocene: Age of Man," *National Geographic*, March, 2011. ngm.nationalgeographic.com.
5. Chelsea Harvey, "CO_2 Levels Just Hit Another Record," *Scientific America*, May 2019 scientificamerican.com.
6. Royal Meteorological Society meeting, "The Pliocene: The Last Time Earth had >400 ppm of Atmospheric CO_2," April 3, 2019, rmets.org.
7. IPCC "Special Report: Global Warming of 1.5°C," 2018, ipcc.ch/sr15/.
8. Franke James, "Franke James is a Climate Mama," 2011, frankejames.com.
9. See the Climate Reality Project website, climaterealityproject.org/.

10. Excerpt from post first published on Elephant Journal website: "Climate Change and Cancer: The Long Goodbye," June 15, 2016, elephantjournal.com.
11. Conference of the Parties (COP) is the decision-making body of the United Nations Framework Convention on Climate Change (UNFCCC). It has met annually since 1995 and is one of the bodies created by the first 1992 United Nations Earth Summit.

Chapter 2

1. Certain shale deposits, like the Marcellus Shale in Pennsylvania, contain radioactive materials, such as uranium and thorium, which decay as radium 226. These are naturally occurring materials, safe when they stay buried underground but unsafe when they are disturbed and/or brought to the surface by fracking.
2. The Climate Accountability Institute engages in research and education on anthropogenic climate change, dangerous interference with the climate system, and the contribution of fossil fuel producers' carbon production to atmospheric carbon dioxide content. See climate accountability.org/.
3. David Coady et al., "Global Fossil Fuel Subsidies Remain Large: An Update Based on Country-Level Estimates," *International Monetary Fund Working Paper*, May, 2019.
4. "The Earth's History in One Year," as quoted by Peter Sinclair on Climate Crocks website, December, 31 2010, climatecrocks.com.
5. Report of the Environmental Pollution Panel, "Restoring the Quality of Our Environment," President's Science Advisory Committee, The White House, November 1965, pp 111–133.
6. Lost to history until 2011, when Eunice Foote's paper was shared widely, her research predates Tyndall's by three years. Gender bias in the mid-18th century hid her from the many lists of key studies that link human fingerprints to our changing climate. More details can be found here: "This Lady Scientist Defined the Greenhouse Effect But Didn't Get the Credit, Because Sexism," by Leila McNeill, December 5, 2016, www.smithsonianmag.com.
7. Benjamin Franta, "On the 100th Birthday of the US Oil Industry in 1959, Edward Teller Warns About Global Warming," *The Guardian* January 1, 2018.
8. William Sellers, "A Global Climatic Model Based on the Energy Balance Of The Earth-Atmosphere System," quoted in *The Warming Papers: The Scientific Foundation for the Climate Change Forecast*, edited by David Archer and Raymond Pierrehumbert, Wiley-Blackwell, 2011.

9. A more detailed background and outline of the work and process of the UNFCCC can be found on its website: UNFCCC.net.
10. Bitumen is a sticky and highly viscous liquid or semi-solid form of petroleum.
11. "Banking on Climate Change," Fossil Fuel Finance Report Card, 2019.
12. The National Academies of Sciences, Engineering, Medicine, "National Academies Presidents Affirm the Scientific Evidence of Climate Change," June 18, 2019.
13. Kate Raworth, "Exploring Doughnut Economics" website: kateraworth.com/doughnut/.
14. Stockholm Resilience Centre, Planetary Boundaries. There are nine boundaries: stratospheric ozone depletion; loss of biosphere integrity (biodiversity loss and extinctions); chemical pollution and the release of novel entities; climate change; ocean acidification; freshwater consumption and the global hydrological cycle; land system change; nitrogen and phosphorus flows to the biosphere and oceans; and atmospheric aerosol loading. See stockholm resilience.org/.
15. The Government of Bhutan's Centre for Bhutan Studies revised and released an updated Gross National Happiness Index in 2011. The Index consists of 33 indicators in nine domains: psychological well-being, health, education, time use, cultural diversity and resilience, good governance, community vitality, ecological diversity and resilience, and living standards. The Index gauges the nation's well-being directly by measuring citizens' achievements in each indicator.
16. OECD (Organisation for Economic Co-operation and Development), *How's Life?* 2017. A statistical report, released every two years, that describes some of the essential aspects of life that shape people's well-being in OECD and partner countries.
17. Nini Shridhar and Lorelei Walker, "Gene-Environment Interactions & Epigenetics," Collaborative on Health and The Environment, August 2016. In addition to this, there are many more studies looking at the intergenerational climate impacts of heat, stress, and air pollution on humans and animals.
18. Climate Justice Alliance: climatejusticealliance.org/just-transition/.
19. US Department of Health and Human Services, Office of Minority Health, "Asthma and African Americans," minorityhealth.hhs.gov/.
20. Robert D. Bullard, Paul Mohai, Robin Saha, Beverly Wright, principal authors. "Toxic Waste and Race at Twenty: 1987–2007," Justice & Witness Ministries, United Church of Christ, March 2007.

Chapter 3

1. EPA, "Assessing and Managing Chemicals under TSCA: The Frank R. Lautenberg Chemical Safety for the 21st Century Act Implementation Activities," epa.gov.
2. The Precautionary Principle states that decisions, policies, and laws should be based on the knowledge that there is the possibility of harm from making a certain decision when extensive scientific knowledge on the matter is unavailable. Society, and by extension our course, should protect the public from exposure to harm, when scientific investigation has found a plausible risk. Different jurisdictions put this into effect in different ways or ignore it entirely. The European Union, for the most part, uses the principle prior to allowing chemicals into the market; the United States, the opposite. Use until proven harmful. This is clear to see when we look at differences in rules with regard to GMOs.
3. *Global Chorus: 365 Voices on the Future of the Planet*, edited by Todd MacLean, Rocky Mountain Books, 2014.
4. Since 2003, many psychologists, professors, and psychiatrists have used the term *solastalgia* to explain the mental health issues associated with depression and concerns related to climate change.
5. Glenn Albrecht, "The Age of Solastalgia," The Conversation website, 2012. theconversation.com.
6. S. Clayton, C. M. Manning, K. Krygsman, and M. Speiser. "Mental Health and Our Changing Climate: Impacts, Implications, and Guidance," Washington, D.C.: American Psychological Association, and ecoAmerica, 2017 ; "*The Lancet* Countdown: Tracking Progress on Health and Climate Change," *The Lancet* 389: 10074, March 18, 2017.
7. Joanna Macy and Chris Johnstone, *Active Hope: How to Face the Mess We're in without Going Crazy*. New World Library, 2012.
8. Kate Davies, *Intrinsic Hope: Living Courageously in Troubled Times*. New Society Publishers, 2018.
9. Jonathan Lear is an American philosopher and professor of philosophy at the University of Chicago. His 2006 book *Radical Hope: Ethics in the Face of Cultural Devastation* discusses this concept in great detail.
10. Published with the permission of the author, Olena Alec. First published on May 12, 2019 on the Climate Reality Project, Reality Hub.
11. Paul Hawken, ed. *Drawdown: The Most Comprehensive Plan Ever Proposed to Reverse Global Warming*. Edited by Paul Hawken, Penguin, 2017.
12. Danielle F. Lawson, "Children Can Foster Climate Change Concern Among Their Parents," *Nature Climate Change* May 2019.
13. Jonathan Watts, "'Biggest Compliment Yet': Greta Thunberg Welcomes Oil Chief's 'Greatest Threat' Label," *The Guardian* July 5, 2019.

Chapter 4

1. The mission of Young Voices for the Planet (YVFP) is to address important issues by revitalizing civic and democracy education. Through uplifting and inspiring success stories, they seek to empower young people to become leaders and take an essential role in informing and empowering themselves, their peers, and their communities. The YVFP project is about creating the fundamental shift in young people's understanding of what's possible and getting them excited about becoming involved in local democratic decision-making. See youngvoicesfortheplanet.com/.
2. As of August 2019, OurKidsClimate.org had 46 member organizations in 19 countries.
3. Lorna Gold, *Climate Generation: Awakening to Our Children's Future*. New City Press, 2019.

Chapter 5

1. *Last Child in the Woods: Saving Our Children from Nature-Deficit Disorder*. Richard Louv, Algonquin Books, 2008.
2. The mission of Young Voices for the Planet (YVFP) is to address important issues by revitalizing civic and democracy education and empowering youth, through uplifting and inspiring success stories, to become leaders and take an essential role in informing and empowering themselves, their peers, and their communities. The YVFP project is about creating the fundamental shift in young people's understanding of what's possible and getting them excited about becoming involved in local democratic decision-making.
3. Our Kids' Climate is a network of climate-parent groups from around the world founded in 2005, who are uniting for climate action on behave of our children. See ourkidsclimate.org/.
4. On September 20, 2019, over 4 million people, led by youth, walked out of school and work, around the world, to protest inaction on the climate crisis.
5. Debi Gliori, *The Trouble with Dragons*. Bloomsbury, 2009.
6. Schools for Climate Action, schoolsforclimateaction.weebly.com/; and National Center for Science Education, ncse.ngo/.
7. Reach out to us at ClimateMama by email, info@climatemama.com, or on Facebook, Instagram, and Twitter @climatemama.

Chapter 6

1. "Global Warming's Six Americas is a well-established segmentation of Americans based on their climate change beliefs, attitudes, and behaviors. The original Six Americas model requires a

36-question-screener, and although there is increasing interest in using these segments to guide education and outreach efforts, the number of survey items required is a deterrent. Using 14 national samples and machine learning algorithms, we identify a subset of four questions from the original 36, the Six Americas Short SurveY (SASSY), that accurately segment survey respondents into the Six Americas categories." Quote from Breanne Chryst, et al., "Global Warming's 'Six Americas Short Survey': Audience Segmentation of Climate Change Views Using a Four Question Instrument," published Aug. 23, 2018 on the Yale Program on Climate Change Communication website: climatecommunication.yale.edu.

2. *Washington Post*, "Washington Post-Kaiser Family Foundation Climate Change Survey, July 9-Aug. 5, 2019," September 16, 2019, washingtonpost.com.
3. Susan Cagle, "'Protesters as terrorists': Growing number of states turn anti-pipeline activism into a crime," *The Guardian*, July 8, 2019.
4. Future Coalition, FutureCoalition.org. The Future Coalition provides youth leaders with the tools, resources, and support to power their ideas and amplify their impacts.

Chapter 7

1. Erica Chenoweth and Maria J. Stephan, *Why Civil Resistance Works*. Columbia University Press, August, 2011, cup.columbia.edu.
2. Jean-Francois Bastin, et al., "The Global Tree Restoration Potential," *Science* Vol 365, Issue 6448, July 5, 2019.
3. Edward Paul Abbey, d. 1989, American author, advocate for public lands, environmentalist, and political commentator.
4. Autumn Peltier, 15-year-old member of the Wiikwemkoong First Nation and Chief Water Commissioner for the Anishinabek Nation. She represented more than 40 First Nations from Ontario, Canada, at the United Nations in New York City during Climate Week, September, 2019.

Climate Action Organizing and Book Club Guide

1. Climate Generation, a Will Steger Legacy, climategen.org/take-action/act-climate-change/climate-stories/discover-climate-story/

Index